Methods in Molecular Biology™

AF773813

Series Editor
John M. Walker
School of Life Sciences
University of Hertfordshire
Hatfield, Hertfordshire, AL10 9AB, UK

For other titles published in this series, go to
www.springer.com/series/7651

Tissue Microarrays

Methods and Protocols

Edited by

Ronald Simon

Institute of Pathology, University Medical Center Hamburg- Eppendorf, Hamburg, Germany

Humana Press

Editor
Dr. Ronald Simon
Institute of Pathology
University Medical Center Hamburg-Eppendorf
Hamburg
Germany
r.simon@uke.uni-hamburg.de

ISSN 1064-3745 e-ISSN 1940-6029
ISBN 978-1-4939-6103-0 ISBN 978-1-60761-806-5 (eBook)
DOI 10.1007/978-1-60761-806-5
Springer New York Dordrecht Heidelberg London

© Springer Science+Business Media, LLC 2010
Softcover re-print of the Hardcover 1st edition 2010
All rights reserved. This work may not be translated or copied in whole or in part without the written permission of the publisher (Humana Press, c/o Springer Science+Business Media, LLC, 233 Spring Street, New York, NY 10013, USA), except for brief excerpts in connection with reviews or scholarly analysis. Use in connection with any form of information storage and retrieval, electronic adaptation, computer software, or by similar or dissimilar methodology now known or hereafter developed is forbidden.
The use in this publication of trade names, trademarks, service marks, and similar terms, even if they are not identified as such, is not to be taken as an expression of opinion as to whether or not they are subject to proprietary rights.
While the advice and information in this book are believed to be true and accurate at the date of going to press, neither the authors nor the editors nor the publisher can accept any legal responsibility for any errors or omissions that may be made. The publisher makes no warranty, express or implied, with respect to the material contained herein.

Printed on acid-free paper

Humana Press is part of Springer Science+Business Media (www.springer.com)

Preface

The tissue microarray (TMA) method presents as a modern high technology, although its roots go back to the 80s when researchers first started to combine several small pieces of tissues into so-called sausage blocks. In this respect, the TMA invention was not firstly characterized by technical improvements, but its true novelty was to link clinical data to the tissues that were combined on one slide. The very high number of tissues that can be included into one TMA, the small size and regular shape of the tissue spots, the preservation of integrity of the donor tissue blocks, and the highly organized array pattern that allows for reliable allocation of clinical data to individual tissue spots made it a discrete technique with unique features.

When the TMA technology was developed 12 years ago, its benefit was controversially debated. While many researchers welcomed the method enthusiastically, there were concerns by others that results obtained from the small tissue cores used for TMA making would not be sufficiently representative of the donor tissues. Meanwhile, the increasing use of this technology has imposingly demonstrated its tremendous utility in research. In fact, basically all clinically relevant associations between molecular markers and clinical endpoints could be reproduced using only one single 0.6 mm core per tissue sample so that TMAs have nowadays become a standard tool allowing for a new dimension of tissue analysis. Rather than just representing an efficient and economic alternative to expensive and time-consuming conventional large section tissue analysis, TMAs enable the analysis of whole populations of tissues with a previously unreached statistical power.

In the era of microarrays that were typically made from spotted cDNAs or oligonucleotides, selection of the term "tissue microarray" for the new emerging tissue analysis technique was probably not optimal. It has led to tremendous efforts to develop automated devices to manufacture, scan, and analyze TMAs rapidly. However, TMAs – in sharp contrast to DNA arrays that comprise homogeneous spots of nucleic acids – represent a miniaturized kind of histopathology. All the critical issues connected with "classical" tissue analysis, e.g., fixation artifacts, antigen retrieval strategies, tissue heterogeneity, differentiation between normal and neoplastic cells, or intra- and interobserver differences in the analysis of specific cytologic structures, are still the same in a TMA than in a conventional large tissue section.

The aim of this book is not only to introduce the world of TMA making and TMA applications, but also to provide insights into the inherent and complex aspects of the most popular assays used for in-situ tissue analysis. Various applications of the TMA technology and the various kinds of sources for TMAs, including, for example, human, animal, and plant tissues; cell lines, or xenografts, make it an attractive technique not only for pathologists, but also for molecular biologists, physicians, and other researchers in the various areas of life sciences.

In addition to those who have contributed to this book, I would like to thank my wife Julia and my daughters Janika and Luka for their patience and understanding for the long hours in the weekends when I was stuck to my laptop.

Hamburg, Germany *Ronald Simon*

Contents

Contributors

VIVIEN ANGELI • *Department of Internal Medicine, Semmelweis University, Budapest, Hungary*
ROBERT L. CAMP • *Department of Pathology, Yale University, New Haven, CT, USA*
ELBERT CHANG • *Department of Anatomy, College of Osteopathic Medicine of the Pacific, Western University of Health Sciences, Pomona, CA, USA*
JASON CHRISTIANSEN • *HistoRx, Inc, New Haven, CT, USA*
FELIX KH CHUN • *Martini-Clinic, Prostate Cancer Center, University Medical Center Hamburg-Eppendorf, Hamburg, Germany*
EDGAR DAHL • *Molecular Oncology Group, Institute of Pathology, University Hospital of the RWTH Aachen, Germany*
MILTON W. DATTA • *Hospital Pathology Associates and Department of Pathology, University of Minnesota, Minneapolis, MN, USA*
MARISA DOLLED-FILHART • *HistoRx, Inc, New Haven, CT, USA*
ANDREAS ERBERSDOBLER • *Department of Pathology, Charité – University Medical Center, Berlin, Germany*
MARLENA SCHOENBERG FEJZO • *Division of Hematology/Oncology, Department of Medicine, School of Medicine, University of California, Los Angeles, Los Angeles, CA, USA*
LEVENTE FICSOR • *Department of Internal Medicine, Semmelweis University, Budapest, Hungary*
MARK GUSTAVSON • *HistoRx, Inc, New Haven, CT, USA*
MELISSA HODEIB • *Department of Anatomy, College of Osteopathic Medicine of the Pacific, Western University of Health Sciences, Pomona, CA, USA*
WILLIAM J. HOWAT • *Histopathology/ISH facility, Li Ka Shing Centre, Cancer Research UK, Cambridge Research Institute, Cambridge, UK*
ZHONGTING HU • *Department of Anatomy, College of Osteopathic Medicine of the Pacific, Western University of Health Sciences, Pomona, CA, USA; NeuBiogene Inc., Research and Development, Pasadena, CA, USA*
ANDRÉ A. KAJDACSY-BALLA • *Department of Pathology, University of Illinois at Chicago, Chicago, IL, USA*
ANNE KALLIONIEMI • *Institute of Medical Technology, University of Tampere and Tampere University Hospital, Tampere, Finland*
TIBOR KRENACS • *Department of Pathology and Experimental Cancer Research, Budapest, Hungary*
JUSTIN LIN • *Department of Cancer Biology, Kimmel Cancer Center, Thomas Jefferson University, Philadelphia, PA, USA*
MARTINA MIRLACHER • *Institute of Pathology, University Medical Center Hamburg Eppendorf, Hamburg, Germany*

BELA MOLNAR • *Department of Internal Medicine, Semmelweis University, Budapest, Hungary*
YASODHA NATKUNAM • *Department of Pathology, Stanford University School of Medicine, Stanford, CA, USA*
ANDRÉA RODRIGUES CORDOVIL PIRES • *Fonte Medicina Diagnóstica Ltda. – Medical Director Universidade Federal Fluminense – Professor Niterói, RJ, Brazil*
SIMONE RABELLO DE SOUZA • *Fonte Medicina Diagnóstica Ltda. – Administrative Director Niterói, RJ, Brazil*
DAVID L. RIMM • *Department of Pathology, Yale University, New Haven, CT, USA*
HALLGEIR RUI • *Department of Cancer Biology, Kimmel Cancer Center, Thomas Jefferson University, Philadelphia, PA, USA*
GUIDO SAUTER • *Institute of Pathology, University Medical Center Hamburg Eppendorf, Hamburg, Germany*
THORSTEN SCHLOMM • *Martini-Clinic, Prostate Cancer Center, University Medical Center Hamburg-Eppendorf, Hamburg, Germany*
RONALD SIMON • *Institute of Pathology, University Medical Center Hamburg Eppendorf, Hamburg, Germany*
ASHLEY BROOKE SJOLUND • *Department of Cancer Biology, Kimmel Cancer Center, Thomas Jefferson University, Philadelphia, PA, USA*
DENNIS J. SLAMON • *Division of Hematology/Oncology, Department of Medicine, School of Medicine, University of California,Los Angeles, Los Angeles, CA, USA*
PIERRE TENNSTEDT • *Institute of Pathology, University Medical Center Hamburg Eppendorf, Hamburg, Germany*
JOHN L. TONKINSON • *HistoRx, Inc, New Haven, CT, USA*
THAI HONG TRAN • *Department of Cancer Biology, Kimmel Cancer Center, Thomas Jefferson University, Philadelphia, PA, USA*
FRANSISCUS ERI UTAMA • *Department of Cancer Biology, Kimmel Cancer Center, Thomas Jefferson University, Philadelphia, PA, USA*
SEBESTYEN VIKTOR VARGA • *Department of Internal Medicine, Semmelweis University, Budapest, Hungary*
JÜRGEN VEECK • *Molecular Oncology Group, Institute of Pathology, University Hospital of the RWTH Aachen, Germany*
SUSAN J. WILSON • *Histochemistry Research Unit, Southampton General Hospital, Southampton, UK*
SHUCHUN ZHAO • *Department of Pathology, Stanford University School of Medicine, Stanford, CA, USA*

Chapter 1

Applications of Tissue Microarray Technology

Ronald Simon

Abstract

Tissue microarrays (TMAs) have become a standard tool for tissue-based research during the last decade. In cancer research, depending on the available data attached to the arrayed tissue, three main types of arrays are commonly manufactured. Prevalence TMAs have no further data available and are suited to estimate the frequency of the occurrence of a particular alteration. Progression arrays include tissues of different stages of disease, and are instrumental to study the role of a marker protein for tumor initiation, progression, or metastatic growth. Prognosis TMAs contain tissues with patient follow-up data. These TMAs are the key to uncover the clinical impact of molecular markers. In combination with normal tissue arrays representing healthy tissues, prevalence, progression, and prognosis TMAs allow for a rapid and comprehensive analysis of molecular markers in human cancers. TMAs are also successfully used for many noncancer applications, such as Alzheimer's or inflammatory disease research.

Key words: Tissue microarray, Prevalence array, Progression array, Prognosis array, Cross reactivity testing, Translational research

1. Introduction

During the last decade, enormous efforts have been made to uncover the structure and function of the human genome. A first draft version of the entire human genome became publicly available in 2001, comprising the nucleotide sequence information of more than 30,000 human genes and RNA transcripts. These data formed the basis for new emerging microarray techniques including cDNA, oligonucleotide, BAC (bacterial artificial chromosome) and protein "chips" allowing for the comprehensive analysis of the entire human genome in a single experiment. These array technologies have been successfully employed to compare expression patterns between diseased and nondiseased tissues in hundreds of published studies, and lead to the identification of a huge number of potentially disease-related genes and proteins. However,

Ronald Simon (ed.), *Tissue Microarrays: Methods and Protocols*, Methods in Molecular Biology, vol. 664,
DOI 10.1007/978-1-60761-806-5_1, © Springer Science+Business Media, LLC 2010

an important limitation of most of these array technologies is that unfixed tissue samples are required. Such tissues are usually collected prospectively from clinical studies or routine surgery, and lack long-term clinical follow-up data. In addition, the high costs of DNA arrays often limit the number of samples that can be included in DNA array-based studies. Consequently, there is a strong need for validation of promising candidate genes. This is optimally done in large sets of clinically and pathologically well-defined tissues, with long-term clinical follow-up information. Such tissues are available from the archives of pathology institutes, but thousands of tissue samples must often be analyzed in studies aiming at a high degree of statistical power. Molecular genetic analysis methods like multi-tissue Northern, Western, or Southern blots, multidimensional gel electrophoresis, or high throughput real-time PCR, together with automated sample processing and liquid handling systems, have been employed for this purpose (1–4). Although these techniques are well suited for high-throughput analyses, they share the disadvantage that tissues must be disintegrated before analysis. This is a considerable drawback because candidate genes can be expressed in multiple different cell types, and it might also be important to distinguish the cellular localization of the gene products. Therefore, in-situ technologies such as immunohistochemistry (IHC), RNA in-situ hybridization (RNA-ISH) or fluorescence in-situ hybridization (FISH) represent optimal means to study molecular epidemiology. However, large-scale in-situ tissue analyses are cumbersome and slow when traditional methods of molecular pathology are used. Also, it is typically not possible to cut more than 200 regular sections from one tissue block. Cutting traditional tissue sections for a high number of in-situ analyses would, therefore, rapidly exhaust valuable tissue resources. The tissue microarray technology overcomes these shortcomings because it allows for simultaneous analysis of hundreds of tissue samples in a single paraffin block.

The range of TMA applications are very broad. In fact, the same types of analysis applicable to conventional large tissue sections can be done in a TMA format. Since the first report on the TMA Technology in 1998, hundreds of publications reviewing or using the TMA approach have been published. Most of them have utilized TMAs in cancer research, but there is a growing body of studies using the TMA technology also for research in non-neoplastic tissues, for example, in inflammatory and neurodegenerative disease. In cancer research, different types of TMAs can be manufactured with an optimized composition of the tissue set and on the dataset attached to the tissues, including prevalence TMAs, progression TMAs, prognostic TMAs, TMA from normal tissues, and TMAs composed of experimental tissues like xenograft tumors or cell lines.

2. Prevalence TMAs

Prevalence TMAs comprise tumor samples from one or several tumor entities without extensive clinico-pathological information. These TMAs are instrumental to estimate the prevalence of a given molecular alteration in tumor entities of interest. Required case numbers are between 30 and 60 samples per tumor type. Prevalence TMAs containing tissue samples from various different tumor entities can be used to comprehensively study the epidemiology of molecular features in human neoplasia. Maximal standardization of TMA experiments allows an effective comparison of frequencies of alteration between different tumor types. The largest "multitumor" TMA manufactured in our laboratory so far contains more than 5,700 different samples from more than 120 different tumor types (5, 6). Such a TMA is not only instrumental to analyze expression patterns of newly detected genes, but can also be of great value for the comprehensive reanalysis even of well-known genes. For example, recent studies have found that the efficacy of Trastuzumab (Herceptin) may not be limited to HER2 positive breast cancers, and suggest that patients with HER2 overexpressing tumors of nonbreast origin also, for example, stomach cancer, might respond to Herceptin (7). Since the success of targeted treatment depends on the presence of the target molecule (Her2) in the cancer cells, comprehensive knowledge of the prevalence of Her2 overexpression or HER2 gene amplification can help to identify all tumor types that might potentially benefit from a particular drug like Herceptin. In a recent study, Tapia et al. (8) analyzed a multitumor TMA with more than 5,700 samples for HER2 amplification by fluorescence in-situ hybridization and for Her2 overexpression by immunohistochemistry. Although Her2 overexpression (score 2/3+) and amplification occurred most often in breast cancers, it was also found in 18 other tumor entities including cancers of the urinary bladder (amplification in 14.3%, overexpression in 6.7%), stomach (8.3/4.9%), endometrium (6.6/6.8%), lung (2.8/3.1%), and ovary (2.3/1.2%, Table 1). Smaller multitumor arrays have been used to study expression of potential new tumor markers, for example, SPANX-B (9), Sil (10), NOX1 (11), COX2, MMP2, or MMP9 (12).

3. Progression TMAs

Progression TMAs are constructed from samples of different stages of one particular tumor type (13–16). We require at least 50 samples per tumor stage/category for statistically relevant analyses.

Table 1
Tumor types with Her2 positivity according to IHC and FISH analysis. IHC/FISH% amplified: Fraction of samples with IHC result 0, 1+, 2+, or 3+ with amplification by FISH, X=no amplification

	IHC				FISH		IHC and FISH analyzable	IHC/FISH % amplified			
Tumor type	*n*	1+(%)	2+(%)	3+(%)	*n*	% amplified	*n*	0	1+	2+	3+
Breast											
Breast, ductal carcinoma	1,466	8.5	3.3	12.3	1,146	20.8	1,124	4.1	12.0	41.7	84.0
Breast, lobular carcinoma	291	3.8	2.4	3.1	243	6.2	216	2.3	9.0	28.6	66.7
Breast, medullary carcinoma	80	6.3	0.0	8.8	78	11.5	69	4.4	X	X	85.7
Breast, mucinous carcinoma	80	10.0	3.8	5.0	66	7.6	61	X	X	33.3	100.0
Breast, apocrine carcinoma	18	5.6	0.0	38.9	15	40.0	15	X	X	X	85.7
Breast, cribriform carcinoma	62	11.3	4.8	3.2	49	10.2	48	4.0	14.3	X	100.0
Breast, papillary carcinoma	27	7.4	3.7	7.4	23	13.0	21	4.5	X	X	100.0
Breast, tubular carcinoma	65	4.6	1.5	0.0	47	4.3	54	1.6	33.3	X	X
Breast, other carcinomas	25	4.0	4.0	16.0	21	9.5	20	X	X	X	50.0
Lung											
Lung, squamous cell carcinoma	44	0.0	0.0	2.3	43	2.3	39	X	X	X	100.0
Lung, adenocarcinoma	42	0.0	0.0	4.8	25	0.0	21	X	X	X	X
Lung, large cell cancer	42	2.4	0.0	2.4	40	5.0	34	X	100.0	X	100.0
Gastrointestinal tract											
Esophagus, adenocarcinoma	7	0.0	0.0	0.0	5	20.0	5	20.0	X	X	X
Esophagus, squamous cell carcinoma	28	3.6	0.0	0.0	25	4.0	21	5.3	0.0	X	X
Stomach, diffuse adenocarcinoma	22	0.0	0.0	4.5	8	12.5	7	X	X	X	100.0
Stomach, intestinal adenocarcinoma	39	5.1	2.6	2.6	28	7.1	23	X	50.0	X	100.0
Colon, adenocarcinoma	41	2.4	0.0	0.0	29	0.0	25	X	0.0	X	X

Gall bladder, adenocarcinoma	21	4.8	0.0	0.0	17	5.9	13	X	100.0	X	X
Pancreas, adenocarcinoma	41	4.9	0.0	0.0	29	6.9	23	4.5	100.0	X	X
Urinary tract											
Urinary bladder cancer, TCC noninvasive (pTa)	22	27.3	4.5	4.5	36	2.8	17	X	0.0	0.0	100.0
Urinary bladder cancer, TCC invasive (pT2-4)	24	0.0	8.3	0.0	34	14.7	14	X	X	50.0	X
Urinary bladder, sarcomatoid cancer	6	16.7	0.0	0.0	8	12.5	6	X	100.0	X	X
Male genital tract											
Prostate cancer, hormone-refractory	33	3.0	3.0	0.0	35	0.0	24	X	X	X	X
Female genital tract											
Ovary, serous cancer	42	2.4	2.4	0.0	40	2.5	33	X	X	100.0	X
Ovary, endometrioid cancer	41	0.0	0.0	0.0	46	2.2	39	2.6	X	X	X
Vulva, squamous cell cancer	32	3.1	0.0	0.0	33	6.1	22	X	100.0	X	X
Endometrium, endometrioid carcinoma	41	0.0	2.4	2.4	44	6.8	26	2.9	X	100.0	100.0
Endometrium, serous carcinoma	18	5.6	0.0	11.1	17	5.9	12	X	0.0	X	100.0
Various tumors											
Pheochromocytoma	25	4.0	0.0	0.0	9	0.0	8	X	0.0	X	X
Glioblastoma multiforme	28	0.0	0.0	0.0	39	2.6	23	X	X	X	X
Fibrosarcoma	7	0.0	0.0	0.0	5	20.0	4	25.0	X	X	X
Skin, benign appendix tumor	22	4.5	0.0	0.0	25	0.0	18	X	X	X	X
PNET	14	0.0	0.0	0.0	14	7.1	14	7.1	X	X	X

With permission from Tapia et al. (8)

Table 2
Composition of a prostate cancer progression TMA

Histological diagnosis	Number of samples
Benign prostatic hyperplasia	65
High-grade PIN	78
Incidental prostate cancer (T1a/b)	95
Localized prostate cancer	180
Hormone-refractory local recurrences	120
Lymph node metastases	20
Distant organ metastases	71

For example, a recently manufactured prostate cancer progression TMA covers 50 samples each from different stages of tumor progression including benign specimens, high-grade prostatic intraepithelial hyperplasia (PIN), untreated localized tumors, local recurrences, nodal, and distant metastases from late-stage hormone-refractory cancers (Table 2). We have also constructed a TMA composed of tissues from 196 nodal positive breast carcinomas. From each tumor, one sample was taken from the primary tumor and from each of three different metastases. Together with samples from 196 nodal negative breast carcinomas, this "breast cancer metastasis TMA" contains almost 1,000 tissue samples. In one study, we used this TMA to demonstrate a high concordance in the HER2 amplification/overexpression between primary tumors and their nodal metastases (17). A multitude of other studies utilized progression TMAs to search for associations between molecular markers and tumor phenotype, for example, cyclin E (16), FGFR1, RAF1 (18), MDM2 or CDK4 (19) amplification or MAGE-A4 expression (20) and stage and grade in bladder cancer, CK7 and CK20 expression and grade in colorectal carcinoma (21), IGFBP2 expression and hormone-refractory state (14), EIF3S3 amplification and stage (22), aneusomy and grade (23) or E-cadherin expression and tumor size (24) in prostate cancer, aneusomy, and tumor type in brain tumors (25), Id-1 expression and metastasis in esophageal cancer (26), CD82 expression and metastasis in breast cancer (27), WT1 and nestin expression in glial tumorigenesis and progression (28), or SHP1 expression and tumor development in lymphomas (29).

4. Prognosis TMAs

Prognosis TMAs contain samples from tumors with available clinical follow-up data. Many early studies used prognosis TMAs to reproduce previously well-established associations between molecular findings and clinical outcome. For example, significant associations were found between estrogen or progesteron expression (30) or HER2 alterations (31) and survival in breast cancer patients, between vimentin expression and prognosis in kidney cancer (32), and between Ki67 labeling index and prognosis in urinary bladder cancer (33), soft tissue sarcoma (34), and in Hurthle cell carcinoma (35). Figure 1 shows the association between HER2 overexpression and amplifications and prognosis in a set of 2,221 breast cancers in a TMA format.

These confirmatory studies provided strong support for the validity and the power of the TMA method. In the following years, the significance of a multitude of new candidate prognosis markers was evaluated with large prognosis TMAs including more than 1,000 tissue samples, for instance, gene amplification and protein expression of HER2, TOP2A, FGFR1, RAF1, MDM2, GLI, CDK4, and E2F3 in urinary bladder cancer (18, 36, 37); Her2, Ki-67, ESR1, hTERT, S6 kinase, EGFR, MYC, CCND1, COX-2, p53, and MDM2 in breast cancer (30, 31, 38–42); p53, MYC, EGFR, HER2, CD117, CD10, IGFBP2, and others in prostate cancer (14, 15, 43–47); as well as numerous additional genes in smaller TMAs of brain (48–50), liver (51), kidney (32), lung (52), prostate (53), breast (54), gastric (54) and colorectal tumors (55–57), Hodgkin's lymphoma (58), and malignant melanoma (59).

Prognosis TMAs are also helpful in determining the optimal diagnostic threshold for a clinical application. Manual analysis of immunohistochemistry typically includes estimation of both the staining intensity and the fraction of stained tumor cells for each tissue spot. A final score is subsequently built from these parameters. For example, spots completely lacking any staining (intensity 0 in 100% of tumor cells) are considered "negative," spots with a 3+ staining intensity in at least 50% of tumor cells are considered "strong," and the spots with staining results in between are

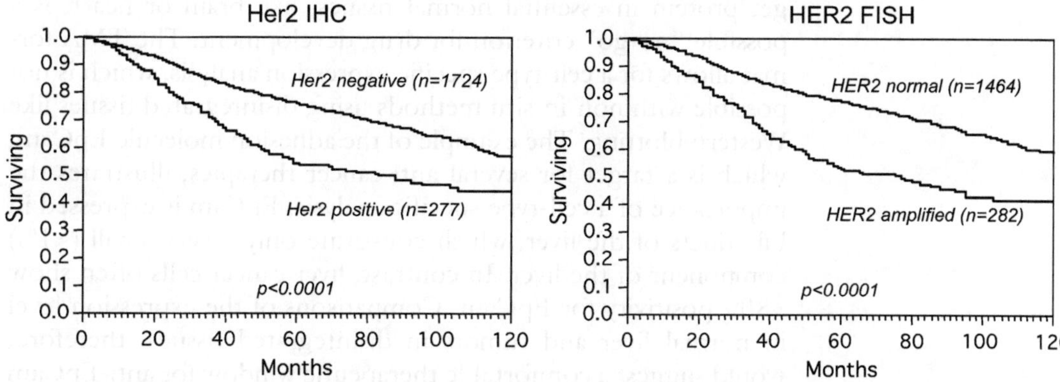

Fig. 1. Kaplan Meier survival plots showing the known association between Her2 expression/amplification in breast cancer. Data from a breast cancer prognosis TMA containing 2,221 breast cancers.

classified in one or two intermediate groups ("weak" or "moderate"). For diagnostic purposes, these groups may then be combined into tumors with "negative" and "positive" staining. Although work intensive, many different thresholds can be compared, and the threshold resulting in the strongest association between target gene expression and patient prognosis will define the best diagnostic cut-off in many instances (Fig. 2).

Prognosis TMAs are typically made from tissues that were retrospectively collected, for instance, from the archives of a pathology institute. These TMAs share the disadvantage that heterogeneously treated patients are included. Future prognosis TMAs will increasingly contain homogeneously treated tumors as clinical trial groups are implementing the construction of TMAs from patients included in clinical trials as part of their protocols. For example, 2,600 breast cancer samples that were analyzed for HER2 amplification and overexpression by the Breast Cancer International Research Group 006 (BCIRG) (60) have been archived in a tissue microarray format and recently been analyzed for TOP2A amplification (Guido Sauter personal communication). This TMA will be highly instrumental to study the impact of new marker proteins on the response to treatment with trastuzumab.

5. Normal Tissue Arrays

Normal tissue arrays are made from healthy tissues. These TMAs are difficult to manufacture, as healthy tissues should normally not be sent to the pathologist. Many commercially available "normal tissue" TMA show significant abnormalities such as inflammatory diseases. Comprehensive knowledge on the expression status of genes in normal tissues is not only important for a better understanding of their role in cancer biology, but is also a crucial issue in drug development. Strong expression of a potential drug target protein in essential normal tissues, like brain or heart, is a possible "no-go" criterion for drug development. The TMA format allows for a cell-type specific expression analysis, which is not possible with non-in-situ methods using disintegrated tissues like Western blotting. The example of the adhesion molecule EpCam, which is a target for several anti-cancer therapies, illustrates the importance of a cell-type specific analysis. EpCam is expressed in bile ducts of the liver, which constitute only a very small (<1%) component of the liver. In contrast, liver cancer cells often show >80% positivity for EpCam. Comparisons of the expression level in normal liver and tumors in disintegrated tissues, therefore, would suggest a comfortable therapeutic window for anti-EpCam drugs. Only the in-situ analysis reveals, however, the high-level

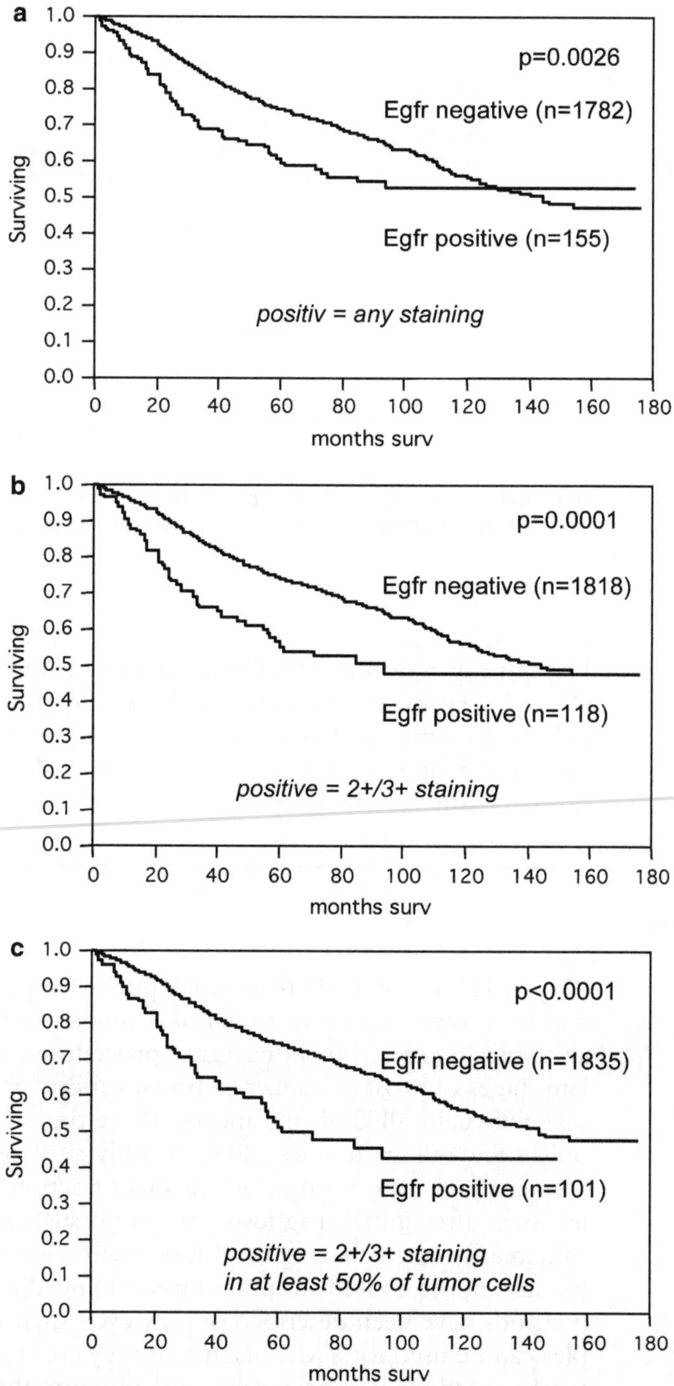

Fig. 2. Optimizing diagnostic thresholds for clinical applications using the example of Egfr immunostaining in a breast cancer prognosis TMA. The strongest prognostic difference between patients with Egfr-positive and Egfr-negative cancers is found if Egfr positivity is defined as 2+/3+ immunostaining intensity in at least 50% of tumor cells (**c**). Other thresholds including any detectable positivity (**a**) or 2+/3+ positivity (as recommended for Her2) (**b**) result in weaker associations.

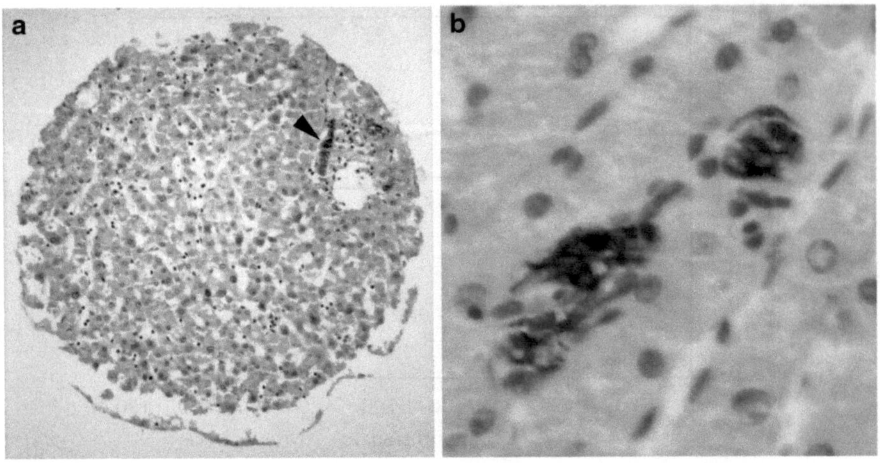

Fig. 3. EpCam expression in normal liver. (**a**) Overview of a TMA spot showing lack of expression in hepatocytes. (**b**) Magnification. High-level EpCam expression is confined to bile ducts (*arrowhead* in **a**). *Microphotograph with permission from Drug Disk Today, Elsevier.*

EpCam expression in a small but vital normal liver compartment (Fig. 3). Human normal tissue TMAs are therefore especially helpful to simulate "drug safety" before entering into clinical phases of drug testing and can be used to facilitate the process of normal tissue reactivity testing.

6. TMAs from Biopsy Specimens

Biopsy TMAs are made from small punch biopsy specimens that had been removed for histological diagnosis of tumor types that are typically treated by nonsurgical procedures, like early or very late stages of prostate cancer or breast cancer, or distant metastases of virtually all kinds of cancers. These tiny samples measuring only a few millimeters are often the only available source for tissues analysis, even though only a small fraction of the sample is left over after initial diagnosis. Although such tissues are highly valuable for research, they cannot be easily used for TMA making because of the limited amount of tissue in needle biopsies. Several methods have been described of how even such small tissue samples can be used for TMA making, if they are stretched, trimmed to the portion of cancer cells, and put upright into the TMA block (61–63). The number of sections that can be obtained from a biopsy TMA obviously depends on the length of the sample. For example, more than 150 sections could be obtained from a prostate cancer biopsy (61). TMAs from biopsy specimens are a paradigm on how the TMA technology can be used for high throughput analysis even of very limited tissue resources.

7. TMAs from Experimental Tissues

TMAs can also be made from experimental tissues like cell lines (19, 64) or xenografts (14). Cell line TMAs containing a large number of different cell lines would be highly useful for the selection of cell lines with distinct genetic features. For example, it would be possible to screen hundreds of arrayed cell lines for expression or copy number changes of a gene of interest (Fig. 4). Suitable cell lines can then be ordered, cultured and, for example, used for functional arrays or serve as model systems for the analysis of particular molecular pathways.

8. TMAs in Alzheimer Research

The use of TMAs is not limited to cancer research. For example, tissue microarrays have also been successfully used for Alzheimer research. In this field of research, the number of amyloid plaques is determined in different areas of the brain. For example, Kellner et al. (65) constructed a large Alzheimer disease (AD) TMA containing 2,325 tissue specimens from three defined regions of 48 AD patients and 48 age-matched controls. The array format is

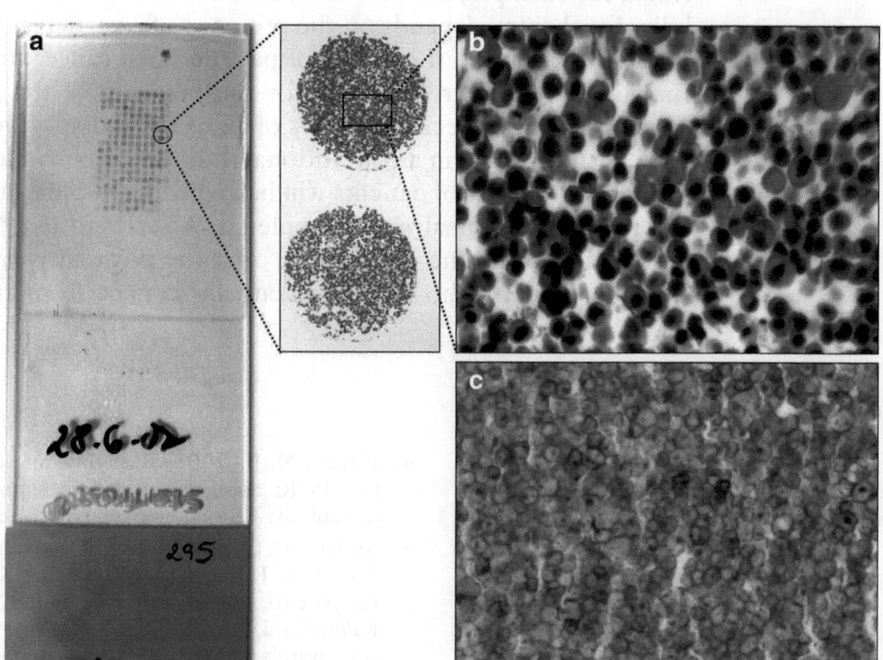

Fig. 4. Cell line TMA containing more than 120 different human cancer cell lines. (**a**) Overview. (**b**) Magnification of a hematoxylin and eosin stained TMA spot. (**c**) Immunostained arrayed cells using an antibody directed against the Egfr.

optimally suited for such kind of analysis because the defined diameter of the tissue spots allows for a highly standardized analysis of areas of exactly the same size. Moreover, multiple samples of all relevant areas of the brain of one patient as well as corresponding control samples from healthy brains can be put into individual array sectors. Alzheimer's disease is also an example where larger size needle diameters such as 1–1.5 mm are better than the cancer-optimized 0.6 mm standard diameters.

9. Other TMA Applications

There are more possible applications of the TMA technology that are likely to become important in the future. For example, TMAs can be used as quality control tools for interlaboratory comparison of IHC performance (66, 67). It is also possible to charge routine diagnosis slides with small TMAs containing a variety of negative and positive controls for commonly used IHC tests (68). Another potentially interesting application is the construction of large and steadily growing TMAs from consecutive routine diagnostic tissue samples. Such a TMA resembles a miniaturized tissue archive of the Pathology institute. "Archive TMAs" might be of great value if a large number of patients need to be quickly reanalyzed, for example, if a new drug becomes available, and if eligibility for therapy depends on the presence of a particular marker. In such a situation, thousands of patients with cancer may wish that their tumors are analyzed for the presence of the therapy target protein. Molecular analysis of all of these tissues would be virtually impossible in a conventional slide-by-slide manner and a significant number of patients will have died from their cancer by the time such an analysis is completed. A ready-to-use "archive TMA" would, on the other hand, offer the possibility to rapidly identify patients who would potentially benefit from the new drug (69).

References

1. Belin, D. (1998) The use of RNA probes for the analysis of gene expression. Northern blot hybridization and ribonuclease protection assay. Methods Mol Biol; 86:87–102.
2. Bichsel, V. E., Liotta, L. A. and Petricoin, E. F., 3rd (2001) Cancer proteomics: from biomarker discovery to signal pathway profiling. Cancer J; 1:69–78.
3. Kallioniemi, O. P. (2001) Biochip technologies in cancer research. Ann Med; 2:142–7.
4. Walker, N. J. (2001) Real-time and quantitative PCR: applications to mechanism-based toxicology. J Biochem Mol Toxicol; 3:121–7.
5. Andersen, C. L., Monni, O., Wagner, U., Kononen, J., Barlund, M., Bucher, C., Haas, P., Nocito, A., Bissig, H., Sauter, G. and Kallioniemi, A. (2002) High-throughput copy number analysis of 17q23 in 3520 tissue specimens by fluorescence in situ hybridization to tissue microarrays. Am J Pathol; 1:73–9.

6. Schraml, P., Bucher, C., Bissig, H., Nocito, A., Haas, P., Wilber, K., Seelig, S., Kononen, J., Mihatsch, M. J., Dirnhofer, S. and Sauter, G. (2003) Cyclin E overexpression and amplification in human tumours. J Pathol; 3:375–82.
7. Gravalos, C. and Jimeno, A. (2008) HER2 in gastric cancer: a new prognostic factor and a novel therapeutic target. Ann Oncol; 9:1523–9.
8. Tapia, C., Glatz, K., Novotny, H., Lugli, A., Horcic, M., Seemayer, C. A., Tornillo, L., Terracciano, L., Spichtin, H., Mirlacher, M., Simon, R. and Sauter, G. (2007) Close association between HER-2 amplification and overexpression in human tumors of non-breast origin. Mod Pathol; 2:192–8.
9. Almanzar, G., Olkhanud, P. B., Bodogai, M., Dell'agnola, C., Baatar, D., Hewitt, S. M., Ghimenton, C., Tummala, M. K., Weeraratna, A. T., Hoek, K. S., Kouprina, N., Larionov, V. and Biragyn, A. (2009) Sperm-derived SPANX-B is a clinically relevant tumor antigen that is expressed in human tumors and readily recognized by human CD4+ and CD8+ T cells. Clin Cancer Res; 6:1954–63.
10. Erez, A., Perelman, M., Hewitt, S. M., Cojacaru, G., Goldberg, I., Shahar, I., Yaron, P., Muler, I., Campaner, S., Amariglio, N., Rechavi, G., Kirsch, I. R., Krupsky, M., Kaminski, N. and Izraeli, S. (2004) Sil overexpression in lung cancer characterizes tumors with increased mitotic activity. Oncogene; 31:5371–7.
11. Geiszt, M., Lekstrom, K., Brenner, S., Hewitt, S. M., Dana, R., Malech, H. L. and Leto, T. L. (2003) NAD(P)H oxidase 1, a product of differentiated colon epithelial cells, can partially replace glycoprotein 91phox in the regulated production of superoxide by phagocytes. J Immunol; 1:299–306.
12. Dicken, B. J., Graham, K., Hamilton, S. M., Andrews, S., Lai, R., Listgarten, J., Jhangri, G. S., Saunders, L. D., Damaraju, S. and Cass, C. (2006) Lymphovascular invasion is associated with poor survival in gastric cancer: an application of gene-expression and tissue array techniques. Ann Surg; 1:64–73.
13. Kononen, J., Bubendorf, L., Kallioniemi, A., Barlund, M., Schraml, P., Leighton, S., Torhorst, J., Mihatsch, M. J., Sauter, G. and Kallioniemi, O. P. (1998) Tissue microarrays for high-throughput molecular profiling of tumor specimens. Nat Med; 7:844–7.
14. Bubendorf, L., Kolmer, M., Kononen, J., Koivisto, P., Mousses, S., Chen, Y., Mahlamaki, E., Schraml, P., Moch, H., Willi, N., Elkahloun, A. G., Pretlow, T. G., Gasser, T. C., Mihatsch, M. J., Sauter, G. and Kallioniemi, O. P. (1999) Hormone therapy failure in human prostate cancer: analysis by complementary DNA and tissue microarrays. J Natl Cancer Inst; 20:1758–64.
15. Bubendorf, L., Kononen, J., Koivisto, P., Schraml, P., Moch, H., Gasser, T. C., Willi, N., Mihatsch, M. J., Sauter, G. and Kallioniemi, O.-P. (1999) Survey of Gene Amplifications during Prostate Cancer Progression by High-Throughput Fluorescence in Situ Hybridization on Tissue Microarrays. Cancer Res; 4:803–6.
16. Richter, J., Wagner, U., Kononen, J., Fijan, A., Bruderer, J., Schmid, U., Ackermann, D., Maurer, R., Alund, G., Knonagel, H., Rist, M., Wilber, K., Anabitarte, M., Hering, F., Hardmeier, T., Schonenberger, A., Flury, R., Jager, P., Fehr, J. L., Schraml, P., Moch, H., Mihatsch, M. J., Gasser, T., Kallioniemi, O. P. and Sauter, G. (2000) High-throughput tissue microarray analysis of cyclin E gene amplification and overexpression in urinary bladder cancer. Am J Pathol; 3:787–94.
17. Simon, R., Nocito, A., Hubscher, T., Bucher, C., Torhorst, J., Schraml, P., Bubendorf, L., Mihatsch, M. M., Moch, H., Wilber, K., Schotzau, A., Kononen, J. and Sauter, G. (2001) Patterns of her-2/neu amplification and overexpression in primary and metastatic breast cancer. J Natl Cancer Inst; 15:1141–6.
18. Simon, R., Richter, J., Wagner, U., Fijan, A., Bruderer, J., Schmid, U., Ackermann, D., Maurer, R., Alund, G., Knonagel, H., Rist, M., Wilber, K., Anabitarte, M., Hering, F., Hardmeier, T., Schonenberger, A., Flury, R., Jager, P., Fehr, J. L., Schraml, P., Moch, H., Mihatsch, M. J., Gasser, T. and Sauter, G. (2001) High-throughout tissue microarray analysis of 3p25 (RAF1) and 8p12 (FGFR1) copy number alterations in urinary bladder cancer. Cancer Res; 11:4514–9.
19. Simon, R., Struckmann, K., Schraml, P., Wagner, U., Forster, T., Moch, H., Fijan, A., Bruderer, J., Wilber, K., Mihatsch, M. J., Gasser, T. and Sauter, G. (2002) Amplification pattern of 12q13-q15 genes (MDM2, CDK4, GLI) in urinary bladder cancer. Oncogene; 16:2476–83.
20. Kocher, T., Zheng, M., Bolli, M., Simon, R., Forster, T., Schultz-Thater, E., Remmel, E., Noppen, C., Schmid, U., Ackermann, D., Mihatsch, M. J., Gasser, T., Heberer, M., Sauter, G. and Spagnoli, G. C. (2002) Prognostic relevance of MAGE-A4 tumor antigen expression in transitional cell carcinoma of the urinary bladder: a tissue microarray study. Int J Cancer; 6:702–5.

21. Park, S. Y., Kim, H. S., Hong, E. K. and Kim, W. H. (2002) Expression of cytokeratins 7 and 20 in primary carcinomas of the stomach and colorectum and their value in the differential diagnosis of metastatic carcinomas to the ovary. Hum Pathol; 11:1078–85.
22. Saramaki, O., Willi, N., Bratt, O., Gasser, T. C., Koivisto, P., Nupponen, N. N., Bubendorf, L. and Visakorpi, T. (2001) Amplification of EIF3S3 gene is associated with advanced stage in prostate cancer. Am J Pathol; 6:2089–94.
23. Skacel, M., Ormsby, A. H., Pettay, J. D., Tsiftsakis, E. K., Liou, L. S., Klein, E. A., Levin, H. S., Zippe, C. D. and Tubbs, R. R. (2001) Aneusomy of chromosomes 7, 8, and 17 and amplification of HER-2/neu and epidermal growth factor receptor in Gleason score 7 prostate carcinoma: a differential fluorescent in situ hybridization study of Gleason pattern 3 and 4 using tissue microarray. Hum Pathol; 12:1392–7.
24. Rubin, M. A., Mucci, N. R., Figurski, J., Fecko, A., Pienta, K. J. and Day, M. L. (2001) E-cadherin expression in prostate cancer: a broad survey using high-density tissue microarray technology. Hum Pathol; 7:690–7.
25. Fuller, C. E., Wang, H., Zhang, W., Fuller, G. N. and Perry, A. (2002) High-throughput molecular profiling of high-grade astrocytomas: the utility of fluorescence in situ hybridization on tissue microarrays (TMA-FISH). J Neuropathol Exp Neurol; 12:1078–84.
26. Li, B., Tsao, S. W., Li, Y. Y., Wang, X., Ling, M. T., Wong, Y. C., He, Q. Y. and Cheung, A. L. (2009) Id-1 promotes tumorigenicity and metastasis of human esophageal cancer cells through activation of PI3K/AKT signaling pathway. Int J Cancer; 125(11):2576–85.
27. Christgen, M., Christgen, H., Heil, C., Krech, T., Langer, F., Kreipe, H. and Lehmann, U. (2009) Expression of KAI1/CD82 in distant metastases from estrogen receptor-negative breast cancer. Cancer Sci; 100(9):1767–71.
28. Rushing, E. J., Sandberg, G. D. and Horkayne-Szakaly, I. (2009) High-grade astrocytomas show increased nestin and Wilms's tumor gene (WT1) protein expression. Int J Surg Pathol.
29. Oka, T., Yoshino, T., Hayashi, K., Ohara, N., Nakanishi, T., Yamaai, Y., Hiraki, A., Sogawa, C. A., Kondo, E., Teramoto, N., Takahashi, K., Tsuchiyama, J. and Akagi, T. (2001) Reduction of hematopoietic cell-specific tyrosine phosphatase SHP-1 gene expression in natural killer cell lymphoma and various types of lymphomas/leukemias: combination analysis with cDNA expression array and tissue microarray. Am J Pathol; 4:1495–505.
30. Torhorst, J., Bucher, C., Kononen, J., Haas, P., Zuber, M., Kochli, O. R., Mross, F., Dieterich, H., Moch, H., Mihatsch, M., Kallioniemi, O. P. and Sauter, G. (2001) Tissue microarrays for rapid linking of molecular changes to clinical endpoints. Am J Pathol; 6:2249–56.
31. Barlund, M., Forozan, F., Kononen, J., Bubendorf, L., Chen, Y., Bittner, M. L., Torhorst, J., Haas, P., Bucher, C., Sauter, G., Kallioniemi, O. P. and Kallioniemi, A. (2000) Detecting activation of ribosomal protein S6 kinase by complementary DNA and tissue microarray analysis. J Natl Cancer Inst; 15:1252–9.
32. Moch, H., Schraml, P., Bubendorf, L., Mirlacher, M., Kononen, J., Gasser, T., Mihatsch, M. J., Kallioniemi, O. P. and Sauter, G. (1999) High-throughput tissue microarray analysis to evaluate genes uncovered by cDNA microarray screening in renal cell carcinoma. Am J Pathol; 4:981–6.
33. Nocito, A., Bubendorf, L., Tinner, E. M., Suess, K., Wagner, U., Forster, T., Kononen, J., Fijan, A., Bruderer, J., Schmid, U., Ackermann, D., Maurer, R., Alund, G., Knonagel, H., Rist, M., Anabitarte, M., Hering, F., Hardmeier, T., Schoenenberger, A. J., Flury, R., Jager, P., Fehr, J. L., Schraml, P., Moch, H., Mihatsch, M. J., Gasser, T. and Sauter, G. (2001) Microarrays of bladder cancer tissue are highly representative of proliferation index and histological grade. J Pathol; 3:349–57.
34. Hoos, A., Stojadinovic, A., Mastorides, S., Urist, M. J., Polsky, D., Di Como, C. J., Brennan, M. F. and Cordon-Cardo, C. (2001) High Ki-67 proliferative index predicts disease specific survival in patients with high-risk soft tissue sarcomas. Cancer; 4:869–74.
35. Hoos, A., Stojadinovic, A., Singh, B., Dudas, M. E., Leung, D. H., Shaha, A. R., Shah, J. P., Brennan, M. F., Cordon-Cardo, C. and Ghossein, R. (2002) Clinical significance of molecular expression profiles of Hurthle cell tumors of the thyroid gland analyzed via tissue microarrays. Am J Pathol; 1:175–83.
36. Simon, R., Atefy, R., Wagner, U., Forster, T., Fijan, A., Bruderer, J., Wilber, K., Mihatsch, M. J., Gasser, T. and Sauter, G. (2003) HER-2 and TOP2A coamplification in urinary bladder cancer. Int J Cancer; 5:764–72.
37. Oeggerli, M., Tomovska, S., Schraml, P., Calvano-Forte, D., Schafroth, S., Simon, R., Gasser, T., Mihatsch, M. J. and Sauter, G. (2004) E2F3 amplification and overexpression

is associated with invasive tumor growth and rapid tumor cell proliferation in urinary bladder cancer. Oncogene; 33:5616–23.

38. Ginestier, C., Charafe-Jauffret, E., Bertucci, F., Eisinger, F., Geneix, J., Bechlian, D., Conte, N., Adelaide, J., Toiron, Y., Nguyen, C., Viens, P., Mozziconacci, M. J., Houlgatte, R., Birnbaum, D. and Jacquemier, J. (2002) Distinct and complementary information provided by use of tissue and DNA microarrays in the study of breast tumor markers. Am J Pathol; 4:1223–33.

39. Poremba, C., Heine, B., Diallo, R., Heinecke, A., Wai, D., Schaefer, K. L., Braun, Y., Schuck, A., Lanvers, C., Bankfalvi, A., Kneif, S., Torhorst, J., Zuber, M., Kochli, O. R., Mross, F., Dieterich, H., Sauter, G., Stein, H., Fogt, F. and Boecker, W. (2002) Telomerase as a prognostic marker in breast cancer: high-throughput tissue microarray analysis of hTERT and hTR. J Pathol; 2:181–9.

40. Ristimaki, A., Sivula, A., Lundin, J., Lundin, M., Salminen, T., Haglund, C., Joensuu, H. and Isola, J. (2002) Prognostic significance of elevated cyclooxygenase-2 expression in breast cancer. Cancer Res; 3:632–5.

41. Al-Kuraya, K., Schraml, P., Torhorst, J., Tapia, C., Zaharieva, B., Novotny, H., Spichtin, H., Maurer, R., Mirlacher, M., Kochli, O., Zuber, M., Dieterich, H., Mross, F., Wilber, K., Simon, R. and Sauter, G. (2004) Prognostic relevance of gene amplifications and coamplifications in breast cancer. Cancer Res; 23:8534–40.

42. Holst, F., Stahl, P. R., Ruiz, C., Hellwinkel, O., Jehan, Z., Wendland, M., Lebeau, A., Terracciano, L., Al-Kuraya, K., Janicke, F., Sauter, G. and Simon, R. (2007) Estrogen receptor alpha (ESR1) gene amplification is frequent in breast cancer. Nature Genetics; 5:655–60.

43. Dhanasekaran, S. M., Barrette, T. R., Ghosh, D., Shah, R., Varambally, S., Kurachi, K., Pienta, K. J., Rubin, M. A. and Chinnaiyan, A. M. (2001) Delineation of prognostic biomarkers in prostate cancer. Nature; 6849:822–6.

44. Mousses, S., Bubendorf, L., Wagner, U., Hostetter, G., Kononen, J., Cornelison, R., Goldberger, N., Elkahloun, A. G., Willi, N., Koivisto, P., Ferhle, W., Raffeld, M., Sauter, G. and Kallioniemi, O. P. (2002) Clinical validation of candidate genes associated with prostate cancer progression in the CWR22 model system using tissue microarrays. Cancer Res; 5:1256–60.

45. Fleischmann, A., Schlomm, T., Huland, H., Kollermann, J., Simon, P., Mirlacher, M., Salomon, G., Chun, F. H., Steuber, T., Simon, R., Sauter, G., Graefen, M. and Erbersdobler, A. (2008) Distinct subcellular expression patterns of neutral endopeptidase (CD10) in prostate cancer predict diverging clinical courses in surgically treated patients. Clin Cancer Res; 23:7838–42.

46. Schlomm, T., Iwers, L., Kirstein, P., Jessen, B., Kollermann, J., Minner, S., Passow-Drolet, A., Mirlacher, M., Milde-Langosch, K., Graefen, M., Haese, A., Steuber, T., Simon, R., Huland, H., Sauter, G. and Erbersdobler, A. (2008) Clinical significance of p53 alterations in surgically treated prostate cancers. Mod Pathol; 21(11):1371–8.

47. Fleischmann, A., Schlomm, T., Kollermann, J., Sekulic, N., Huland, H., Mirlacher, M., Sauter, G., Simon, R. and Erbersdobler, A. (2009) Immunological microenvironment in prostate cancer: High mast cell densities are associated with favorable tumor characteristics and good prognosis. Prostate; 69(9):976–81.

48. Miettinen, H. E., Jarvinen, T. A., Kellner, U., Kauraniemi, P., Parwaresch, R., Rantala, I., Kalimo, H., Paljarvi, L., Isola, J. and Haapasalo, H. (2000) High topoisomerase IIalpha expression associates with high proliferation rate and and poor prognosis in oligodendrogliomas. Neuropathol Appl Neurobiol; 6:504–12.

49. Sallinen, S. L., Sallinen, P. K., Haapasalo, H. K., Helin, H. J., Helen, P. T., Schraml, P., Kallioniemi, O. P. and Kononen, J. (2000) Identification of differentially expressed genes in human gliomas by DNA microarray and tissue chip techniques. Cancer Res; 23: 6617–22.

50. Miettinen, H. E., Paunu, N., Rantala, I., Kalimo, H., Paljarvi, L., Helin, H. and Haapasalo, H. (2001) Cell cycle regulators (p21, p53, pRb) in oligodendrocytic tumors: a study by novel tumor microarray technique. J Neurooncol; 1:29–37.

51. Wang, Y., Wu, M. C., Sham, J. S., Zhang, W., Wu, W. Q. and Guan, X. Y. (2002) Prognostic significance of c-myc and AIB1 amplification in hepatocellular carcinoma. A broad survey using high-throughput tissue microarray. Cancer; 11:2346–52.

52. Brannan, J. M., Dong, W., Prudkin, L., Behrens, C., Lotan, R., Bekele, B. N., Wistuba, I. and Johnson, F. M. (2009) Expression of the receptor tyrosine kinase EphA2 is increased in smokers and predicts poor survival in non-small cell lung cancer. Clin Cancer Res; 13:4423–30.

53. Li, R., Erdamar, S., Dai, H., Sayeeduddin, M., Frolov, A., Wheeler, T. M. and Ayala, G. E.

(2009) Cytoplasmic accumulation of glycogen synthase kinase-3beta is associated with aggressive clinicopathological features in human prostate cancer. Anticancer Res; 6:2077–81.

54. McSherry, E. A., McGee, S. F., Jirstrom, K., Doyle, E. M., Brennan, D. J., Landberg, G., Dervan, P. A., Hopkins, A. M. and Gallagher, W. M. (2009) JAM-A expression positively correlates with poor prognosis in breast cancer patients. Int J Cancer; 6:1343–51.
55. Chung, G. G., Provost, E., Kielhorn, E. P., Charette, L. A., Smith, B. L. and Rimm, D. L. (2001) Tissue microarray analysis of beta-catenin in colorectal cancer shows nuclear phospho-beta-catenin is associated with a better prognosis. Clin Cancer Res; 12:4013–20.
56. Otsuka, M., Kato, M., Yoshikawa, T., Chen, H., Brown, E. J., Masuho, Y., Omata, M. and Seki, N. (2001) Differential expression of the L-plastin gene in human colorectal cancer progression and metastasis. Biochem Biophys Res Commun; 4:876–81.
57. Hoos, A., Nissan, A., Stojadinovic, A., Shia, J., Hedvat, C. V., Leung, D. H., Paty, P. B., Klimstra, D., Cordon-Cardo, C. and Wong, W. D. (2002) Tissue microarray molecular profiling of early, node-negative adenocarcinoma of the rectum: a comprehensive analysis. Clin Cancer Res; 12:3841–9.
58. Garcia, J. F., Camacho, F. I., Morente, M., Fraga, M., Montalban, C., Alvaro, T., Bellas, C., Castano, A., Diez, A., Flores, T., Martin, C., Martinez, M. A., Mazorra, F., Menarguez, J., Mestre, M. J., Mollejo, M., Saez, A. I., Sanchez, L. and Piris, M. A. (2003) Hodgkin and Reed-Sternberg cells harbor alterations in the major tumor suppressor pathways and cell-cycle checkpoints: analyses using tissue microarrays. Blood; 2:681–9.
59. Kielhorn, E., Provost, E., Olsen, D., D'Aquila, T. G., Smith, B. L., Camp, R. L. and Rimm, D. L. (2003) Tissue microarray-based analysis shows phospho-beta-catenin expression in malignant melanoma is associated with poor outcome. Int J Cancer; 5:652–6.
60. Press, M. F., Sauter, G., Bernstein, L., Villalobos, I. E., Mirlacher, M., Zhou, J. Y., Wardeh, R., Li, Y. T., Guzman, R., Ma, Y., Sullivan-Halley, J., Santiago, A., Park, J. M., Riva, A. and Slamon, D. J. (2005) Diagnostic evaluation of HER-2 as a molecular target: an assessment of accuracy and reproducibility of laboratory testing in large, prospective, randomized clinical trials. Clin Cancer Res; 18:6598–607.
61. Datta, M. W., Kahler, A., Macias, V., Brodzeller, T. and Kajdacsy-Balla, A. (2005) A simple inexpensive method for the production of tissue microarrays from needle biopsy specimens: examples with prostate cancer. Appl Immunohistochem Mol Morphol; 1:96–103.
62. Obermann, E. C., Marienhagen, J., Stoehr, R., Wuensch, P. H. and Hofstaedter, F. (2005) Tissue microarray construction from bone marrow biopsies. Biotechniques; 39(6):822, 4, 6.
63. Jhavar, S., Bartlett, J., Kovacs, G., Corbishley, C., Dearnaley, D., Eeles, R., Khoo, V., Huddart, R., Horwich, A., Thompson, A., Norman, A., Brewer, D., Cooper, C. S. and Parker, C. (2009) Biopsy tissue microarray study of Ki-67 expression in untreated, localized prostate cancer managed by active surveillance. Prostate Cancer Prostatic Dis; 2:143–7.
64. Hoos, A. and Cordon-Cardo, C. (2001) Tissue microarray profiling of cancer specimens and cell lines: opportunities and limitations. Lab Invest; 10:1331–8.
65. Kellner, A., Matschke, J., Bernreuther, C., Moch, H., Ferrer, I. and Glatzel, M. (2009) Autoantibodies against beta-amyloid are common in Alzheimer's disease and help control plaque burden. Ann Neurol; 1:24–31.
66. Mengel, M., von Wasielewski, R., Wiese, B., Rudiger, T., Muller-Hermelink, H. K. and Kreipe, H. (2002) Inter-laboratory and interobserver reproducibility of immunohistochemical assessment of the Ki-67 labelling index in a large multi-centre trial. J Pathol; 3:292–9.
67. von Wasielewski, R., Mengel, M., Wiese, B., Rudiger, T., Muller-Hermelink, H. K. and Kreipe, H. (2002) Tissue array technology for testing interlaboratory and interobserver reproducibility of immunohistochemical estrogen receptor analysis in a large multicenter trial. Am J Clin Pathol; 5:675–82.
68. Packeisen, J., Buerger, H., Krech, R. and Boecker, W. (2002) Tissue microarrays: a new approach for quality control in immunohistochemistry. J Clin Pathol; 8:613–5.
69. Sauter, G. and Mirlacher, M. (2002) Tissue microarrays for predictive molecular pathology. J Clin Pathol; 8:575–6.

Chapter 2

Quality Aspects of TMA Analysis

Pierre Tennstedt and Guido Sauter

Abstract

The quality of a TMA experiment can only be as good as the worst step in the entire process. Making and analyzing of tissue microarrays (TMAs) is a complex process involving tissue sample acquisition and preparation, expert histological diagnosis, long-term collection of clinical data, development of suitable laboratory protocols, the microscopic analysis, and the statistical analysis of the experiment. The quality of all these factors equally influences the outcome of TMA studies.

Key words: Tissue microarray, Formalin fixation, Antibody, Tissue preservation

1. Quality of the Tissue

In case of tissue ischemia, degradation of DNA, RNA, and proteins starts almost immediately. It is important to understand that ischemia starts during surgery as soon as the blood supply of an organ or a segment of an organ has been interrupted. The time span between ischemia and removal of a tissue from the body is highly variable and depends on the surgeon, the type of surgery, and on many external factors such as the time required for frozen section analysis of resection borders. Only when the tissue is removed from the body, strategies can be applied to minimize any further degree of DNA, RNA, and protein degradation. Such strategies include freezing in liquid nitrogen, immersion in special agents to stabilize nucleic acids, or fixation of the proteins to allow for best possible preservation of the tissue architecture. A short delay time between surgery and tissue conservation will undoubtedly help to have an acceptable tissue quality. Unfortunately, there is evidence that most of the RNA degradation occurs during surgery and that the more standardizable postsurgery time has much less relevance (1). These data basically demonstrate that

Ronald Simon (ed.), *Tissue Microarrays: Methods and Protocols*, Methods in Molecular Biology, vol. 664,
DOI 10.1007/978-1-60761-806-5_2, © Springer Science+Business Media, LLC 2010

uncontrolled intrasurgical tissue decay greatly limits all attempts for standardized tissue banking.

It is unfortunate, that postsurgical tissue fixation also cannot be fully standardized. Formalin fixation causes proteins to cross-link and makes them resistent against microbial degradation or autolysis. The efficiency of the fixation process, that is, the degree of protein cross-linking, depends on the proper penetration of the formalin into the tissue. The average penetration time of formalin is 1 mm per hour (2). The speed of penetration by a fixative is greatly dependent on the size and the composition of a given tissue. For example, great amounts of fat and presence of dense fibrous tissue hinder formalin penetration. If a large piece of tissue is put in the fixative, proteins in the peripheral areas of the tissue may become more efficiently fixed than the proteins in the inner parts. Consequently, RNA and protein decay will also continue for a longer time in the inner parts of a tissue lump. Very often, some pieces of a tissue are hardly penetrated by formalin. This is especially true if the fixation time is short as is often the case under diagnostic time pressure. Such poorly fixed areas will undergo ethanol fixation during subsequent tissue processing and thus respond differently to many antibodies than formalin-fixed structures.

We believe that tissue quality and formalin fixation can hardly be standardized. Moreover, the most valuable tissue samples with long-term follow-up data have been stored for years in the pathology archives. For these valuable tissues, a standardization of tissue acquisition and fixation has not even been attempted. These problems are obviously not TMA specific; they also concern conventional large tissue sections. While many groups fight a Don Quichotte type of battle with standardization, accepting some "tissue problems" represents a promising alternative approach. It is our opinion that the available TMA data demonstrate that the large number of samples included into TMA studies easily compensate for some tissue shortcomings. All previously established associations between clinical parameters and molecular features were confirmed in TMA studies (for instance HER2 breast (3, 4), ER, PR, p53 (5), Ki67 (6)).

Two practical issues remain to be considered in terms of tissue quality. First, tissue processing may vary significantly between different centers. Combining tissues from different sources is therefore always a risk. This is especially the case if tissues from different sources are compared with each other. For example, in several studies we compared Saudi versus Swiss tumors. All these analyses were limited to FISH analyses because we were concerned about different tissue processing procedures leading to systematic differences in IHC staining (7). Second, it is sometimes recommended to test the suitability of tissues for IHC by analysis of proteins like vimentin, cytokeratins, or transferrin-receptor (8) before a TMA is made. We do not use such approaches. Considering that degradation and fixation effects may not be the same for all proteins and that typical "test" proteins are selected because of their ubiquitous and stable expression, it does not seem too likely that the ability to detect these "easy" proteins can also predict the ability to detect more fragile and sensible proteins. Once a TMA is made, we believe that the applicability of FISH (4) and strong Ki67 (MIB1) staining in some cells and mitoses represents the best available test criteria. MIB1, which strongly stains in mitoses (9), often stains weakly in sub optimally processed tissues.

2. Quality of the Histological Diagnosis

The construction of a large TMA usually requires combining donor samples that have been collected over a long period of time, i.e., 5–10 years. To select suitable cases, the pathology database is screened for the diagnosis of interest. It may be, however, that the diagnostic criteria have changed over the years, or that different pathologists with different diagnostic skills were involved so that the histological diagnosis in the patient's files may be biased. There are many examples in the literature showing that interobserver differences are a common phenomenon in pathology institutes. For example, the histological distinction between non-invasive (pTa) and minimally invasive (pT1) tumors of the urinary bladder is particularly difficult. Tosoni et al. (10) reviewed the histological sections of 235 bladder tumors that were originally considered pT1, and re-classified 35% as pTa, 56% as pT1, 6% as pT1- (at least pT1), and 3% as pT2-4 so that the initial diagnosis was confirmed in only 44% of cases. In 39% of all biopsies, there were also interobserver differences in tumor grade. In order to avoid such a potential bias because of variations in the histological classification, it is crucial that one expert pathologist carefully revises all sections of the candidate tissues before the TMA is made. This is also often needed because classification schemes change over time and should be adjusted to modern standards. As a rule of thumb, the review of all tumors included into one TMA by one pathologist is an important quality criterion for a TMA. This is all the more true, if this pathologist has a documented expertise in the particular field.

3. Quality of Clinical and Pathological Data

The kind and amount of data attached to arrayed tissue samples ultimately determine the options of TMA studies. For example, if no pathological or clinical data are available, it will only be possible to draw conclusions on the prevalence of a particular molecular alteration within the analyzed tissue set. In most cases, 50–100 tissue samples will be sufficient to obtain this result. The scope of most TMA studies, however, especially in the field of cancer research, is to find associations between molecular changes and tissue (tumor) phenotype, that is, histo-pathological stage and grade of cancers, presence of metastases, or patient prognosis. For such a study, the most important data may include pathological data like staging (pT) and grading (G), nodal stage (pN), presence of distant metastases (pM), and histological tumor subtype (e.g., adeno-, squamous cell-, small cell cancer etc.), or clinical data like patient age, gender, previous and subsequent therapy information (surgery, radiation, cytotoxic treatment), and success of the surgery (tumor-free resection margins or not). In addition, there may be tumor type-specific data that are important to address specific questions, for example, the hormone receptor status (estrogen and progesterone receptor) of breast cancer, the preoperative PSA serum level (prostate specific antigen) in case of prostate cancer, tumor localization of colon cancer (left- vs. right-sided), or the kind and success of targeted treatment (e.g., Herceptin and breast cancer, Iressa or Tarceva and lung cancer, Erbitux and colon cancer). Some of these data can be taken from the patient's files. Clinical follow-up data allow for the best level of study results. For statistical analysis, such data must include a time interval (typically in months) between surgery and the clinical endpoint (the "event"; e.g., patient death, tumor progression, or tumor recurrence), and a dichotomous censor (typically 0 or 1) that determines whether the event has happened or not. For patient survival/death, the censor can be "raw" or "tumor specific." The "tumor-specific" censor is used to indicate that the patient has actually died from cancer after the designated period of time, while the "raw" censor only states that the patient has died, no matter for what reason.

Whether tumor-specific survival data are more valuable than raw survival data if the prognostic impact of a particular molecular alteration is to be analyzed is disputed. Assigning the cause of death of a patient to a cancer is often difficult and subjective. Did a cachectic patient with two liver metastases of 1–2 cm each and one cervical lymph node metastasis of 1-cm diameter die from colon cancer or not? Some authors therefore prefer raw survival data as these are at least not biased by subjective interpretations. The associations found between classical histological features (grade and stage) and clinical outcome are the best parameter for the quality of the clinical data set. Finally, it is important to note that the use of patient data may be restricted by ethical regulations that may vary in different countries. These regulations should be strictly followed and considered already in the planning phase of TMA making.

4. Quality of the Experimental Procedure

The vast majority of published TMA analyses have been performed by immunohistochemical analysis of TMAs made from formalin-fixed tissue samples. Immunohistochemistry (IHC), however, is a difficult technique the outcome of which is influenced by a variety of factors including choice of the antibody, choice of experimental conditions and antigen-retrieval strategies, and the condition of the tissue itself. For example, more than 1,500 studies have analyzed the expression patterns of the epidermal growth factor receptor (Egfr) protein in different tumor types by means of immunohistochemistry to date. In these studies, the frequency of Egfr positivity ranges between for example 14 and 53.5% in breast cancer (11, 12), 27 and 87% in urinary bladder cancer (13, 14), 28 and 95% in ovarian cancer (15, 16), and 28 and 87% in colon cancer (17, 18). While the TMA analysis offers maximal experimental standardization (all tumors are stained on 1 day in one set of reagents), it is of critical importance to select the best possible experimental conditions. For this purpose, it is often helpful to select IHC conditions that can distinguish protein expression levels at relevant thresholds, for example, defined by certain tumors or cell lines that can then be used for protocol optimization.

Finding the "optimal" experimental conditions is particularly difficult for new markers where no or little information is available for suitable positive or negative control tissues or the expected intracellular localization of the target protein. In such a case, we prefer experimental conditions resulting in the best obtainable dynamic range of immunostaining intensities. In the optimal case, a staining will result in some normal cell types or tumor cells with very strong positivity, while others remain without any detectable staining.

Another important point to consider is the condition of the TMA section used for IHC. Several studies have suggested that the IHC performance of tissues gradually decreases once a section has been taken from the paraffin block and mounted on the slide (19–21). The same holds also true for TMA sections. In our own study, we compared the IHC results of different standard marker proteins obtained from freshly cut TMA sections as compared to sections that were stored for 6 months at 4°C (22). It showed that the frequency of positivity on old sections decreased from 65 to 46% for ER ($P<0.0001$), from 33 to 18.5% for PR ($P<0.0001$), from 16.3 to 9.6% for HER2 ($P=0.0047$), from 45.1 to 37.7% for cyclin D1 ($P=0.10$), and from 58.9 to 32.9% for E-cadherin ($P<0.0001$). However, this study was also an example for the suitability of TMA studies to detect clinicopathological associations, because despite the lower fraction of positive cases, most associations between IHC data and tumor phenotype that were observed in fresh section analysis were still found when old section data were analyzed. As a result of these studies, we always use sections that are less than 2-weeks-old for TMA analysis. The same quality criteria apply for sections to be used for protocol development.

The impact of experimental conditions on TMA studies is nicely illustrated by a recent own study on p53 in prostate cancer (23). In more than 2,500 early stage primary prostate cancers, we found 2.5% p53 positive cases by IHC. This number corresponded well with sequencing data from almost 100 selected tumors. In addition, p53 immunostaining was strongly associated with tumor recurrence. One of the reviewers of this manuscript repeatedly doubted that the fraction of p53 positive cases is more influenced by the IHC protocol than by tumor heterogeneity. These recurrent complaints forced us to prove our point. We thus modified our protocol using a 180-fold higher antibody concentration and demonstrated that we were able to find – on request – more than 90% p53 positive prostate cancers (Fig. 1). The manuscript was then promptly accepted (23).

Similar problems also arise with RNA in-situ hybridization (RISH) expression analysis on TMA sections, where the sensitivity of the probe has a similar impact on the outcome of the study as the quality of the antibody in an IHC experiment. "Riboprobes" made from cRNA are technically challenging but offer a significantly increased sensitivity as compared to "simple" DNA probes made by asymmetric PCR. In addition, radioactive labels are more sensitive then brightfield dyes (24). Another important problem of RISH is the quality of the RNA in the tissue sample. Particularly in formalin fixed tissues, there will always be a certain fraction of false negative results due to insufficient RNA quality.

DNA copy number analysis using fluorescence in-situ hybridization (FISH) has shown a high degree of concordance between TMA studies performed in different labs (3, 25). In contrast to IHC or RISH, FISH is based on counting of individual fluorescence signals rather than on estimation of staining intensities. This makes the FISH analysis largely independent of the signal intensity. Experimental variations do not exert a strong influence on the study outcome, because FISH signals – even if they are only weak – can be reliably counted as long as they are clearly visible. In addition, FISH is not prone to false negative results like IHC or RISH. Lack of FISH signals indicates that the assay has failed, whereas a "true" negative result is characterized by the normal copy number state in FISH analysis, that is, two FISH signals in the cell nucleus.

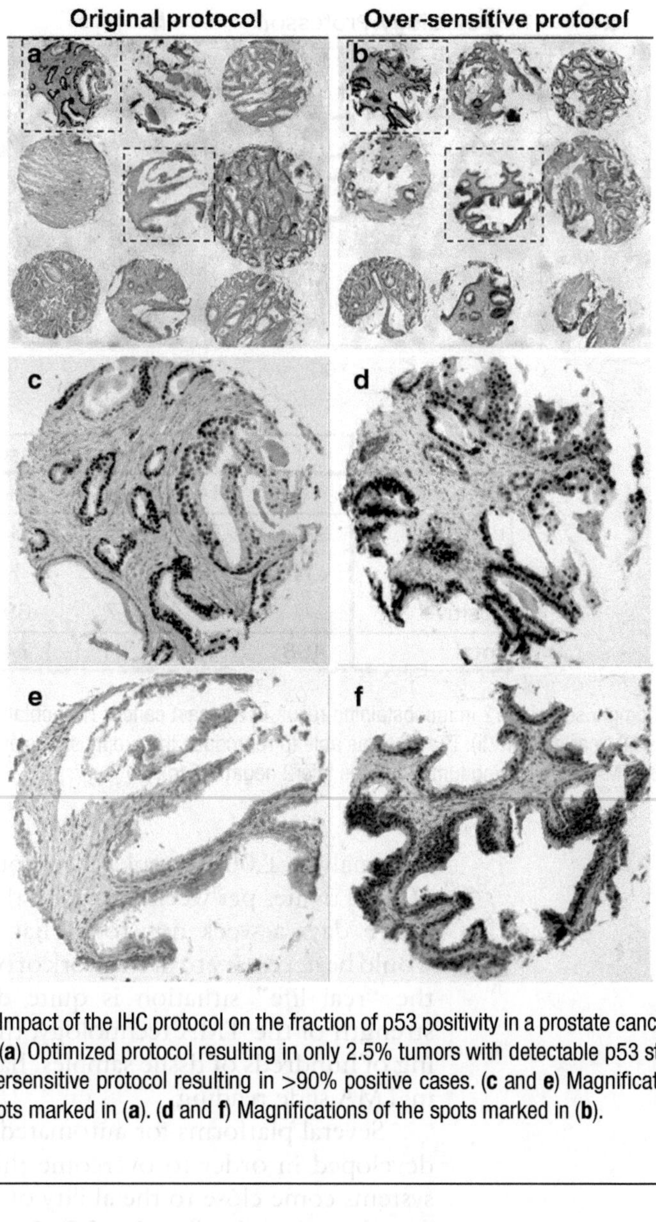

Fig. 1. Impact of the IHC protocol on the fraction of p53 positivity in a prostate cancer TMA study. (**a**) Optimized protocol resulting in only 2.5% tumors with detectable p53 staining. (**b**) Oversensitive protocol resulting in >90% positive cases. (**c** and **e**) Magnifications of the spots marked in (**a**). (**d** and **f**) Magnifications of the spots marked in (**b**).

5. Quality of the Analysis

Human tissue represents a complex mixture of different cell types and extracellular components. Even in cancer samples, there will be not only tumor cells, but also stroma cells, infiltrating lymphoctes, blood vessels, muscle, and other non-neoplastic cells. Tissue analysis, therefore, principally requires pathology skills. A trained pathologist is able to distinguish positivity in different subcellular compartments as well as to exclude non-neoplastic tissues from his analysis. However, a single technician can easily stain 200,000 tissue samples per week using TMAs, but even a skilled pathologist who

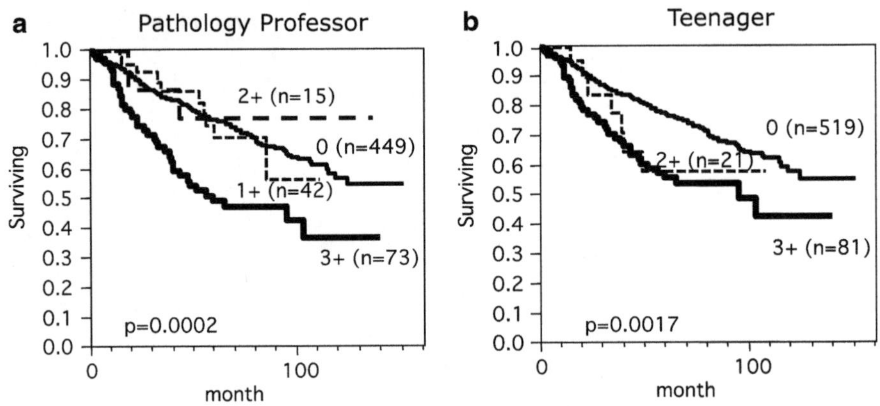

c

	Pathology Professor				
Teenager	0	1+	2+	3+	no tumor
negative	440	32	5	3	35
weak	10	6	3	1	1
positive	0	4	7	69	
total	468	42	17	74	44

Fig. 2. Comparison of Her2 immunostaining result in a breast cancer TMA obtained from a pathology professor (**a**) and from his 10-year-old son (**b**). The boy was able to reproduce the prognostic relevance of Her2 even though he misdiagnosed 35 TMA spots lacking tumor cells as "Her2 negative" (**c**).

can analyze 1,000 samples per hour will only manage to read 40,000 tissues per week, provided the pathologist works 8 h a day and 5 days a week nonstop. That means that five pathologists would be necessary to fit the work of one technician. Unfortunately, the "real life" situation is quite different from this, and the strength of the TMA technology, that is, the simultaneous staining of hundreds of tissue samples, has generated a new bottleneck in TMA slide reading.

Several platforms for automated analysis of TMAs have been developed in order to overcome this bottleneck. Some of these systems come close to the ability of a trained pathologist, at least for selected antibodies (26, 27). However, the requirements vary substantially between different antibodies/targets. In order to demonstrate how easy TMA reading can be, we had a pathology professor and his 10-year-old son to read the same Her2 stained TMA containing 668 breast cancer samples. As expected, the professor's scoring according to Her2 routine diagnostic guidelines (0, 1+, 2+, 3+) resulted in 15% of Her2 2+/3+ positive breast cancers, which had a worse prognosis as compared to the Her2 negative (0/1+) tumors. The 10-year-old boy had no previous training in tissue analysis or IHC scoring, and, therefore, could not use the routine diagnostic scoring system. However, after a brief introduction into microscope operation, he was asked to decide whether there was brown staining in a tissue spot or not, and – if possible – to distinguish between weak and strong staining. Surprisingly, the boy's reading resulted in >90% concordant results with the professor's analysis and also in all expected associations with grade and stage. Also both interpreters came to the same conclusions with respect to the prognostic impact of Her2 overexpression in breast cancer (Fig. 2). This study also shows that large-scale TMAs represent a very robust research tool. Even in case of interpretation by a novice and despite the fact that 36 tissues not containing tumor were falsely included in the boy's analysis, clear-cut associations with prognosis could be demonstrated.

References

1. Schlomm, T., Nakel, E., Lubke, A., Buness, A., Chun, F. K., Steuber, T., Graefen, M., Simon, R., Sauter, G., Poustka, A., Huland, H., Erbersdobler, A., Sultmann, H., Hellwinkel, O. J. (2007) Marked gene transcript level alterations occur early during radical prostatectomy. Eur Urol; 53(2):333–44.
2. Fox, C. H., Johnson, F. B., Whiting, J., Roller, P. P. (1985) Formaldehyde fixation. J Histochem Cytochem; 8:845–53.
3. Al-Kuraya, K., Schraml, P., Torhorst, J., Tapia, C., Zaharieva, B., Novotny, H., Spichtin, H., Maurer, R., Mirlacher, M., Kochli, O., Zuber, M., Dieterich, H., Mross, F., Wilber, K., Simon, R., Sauter, G. (2004) Prognostic relevance of gene amplifications and coamplifications in breast cancer. Cancer Res; 23:8534–40.
4. Tapia, C., Schraml, P., Simon, R., Al-Kuraya, K. S., Maurer, R., Mirlacher, M., Novotny, H., Spichtin, H., Mihatsch, M. J., Sauter, G. (2004) HER2 analysis in breast cancer: reduced immunoreactivity in FISH non-informative cancer biopsies. Int J Oncol; 6:1551–7.
5. Torhorst, J., Bucher, C., Kononen, J., Haas, P., Zuber, M., Kochli, O. R., Mross, F., Dieterich, H., Moch, H., Mihatsch, M., Kallioniemi, O. P., Sauter, G. (2001) Tissue microarrays for rapid linking of molecular changes to clinical endpoints. Am J Pathol; 6:2249–56.
6. Ruiz, C., Seibt, S., Al Kuraya, K., Siraj, A. K., Mirlacher, M., Schraml, P., Maurer, R., Spichtin, H., Torhorst, J., Popovska, S., Simon, R., Sauter, G. (2006) Tissue microarrays for comparing molecular features with proliferation activity in breast cancer. Int J Cancer; 9:2190–4.
7. Al-Kuraya, K., Schraml, P., Sheikh, S., Amr, S., Torhorst, J., Tapia, C., Novotny, H., Spichtin, H., Maurer, R., Mirlacher, M., Simon, R., Sauter, G. (2005) Predominance of high-grade pathway in breast cancer development of Middle East women. Mod Pathol; 7:891–7.
8. FDA (1997) U.S. Food and Drug Administration (FDA). In vitro diagnostic devices: guidance for the preparation of 510(k) submissions – Appendix K – points to consider for review of calibration and quality control labeling for in vitro diagnostic devices. 2/1/96; 1–50.
9. Gerdes, J., Lemke, H., Baisch, H., Wacker, H. H., Schwab, U., Stein, H. (1984) Cell cycle analysis of a cell proliferation-associated human nuclear antigen defined by the monoclonal antibody Ki-67. J Immunol; 4:1710–5.
10. Tosoni, I., Wagner, U., Sauter, G., Egloff, M., Knonagel, H., Alund, G., Bannwart, F., Mihatsch, M. J., Gasser, T. C., Maurer, R. (2000) Clinical significance of interobserver differences in the staging and grading of superficial bladder cancer. BJU Int; 1:48–53.
11. Wrba, F., Reiner, A., Ritzinger, E., Holzner, J. H., Reiner, G. (1988) Expression of epidermal growth factor receptors (EGFR) on breast carcinomas in relation to growth fractions, estrogen receptor status and morphological criteria. An immunohistochemical study. Pathol Res Pract; 1:25–9.
12. Bhargava, R., Gerald, W. L., Li, A. R., Pan, Q., Lal, P., Ladanyi, M., Chen, B. (2005) EGFR gene amplification in breast cancer: correlation with epidermal growth factor receptor mRNA and protein expression and HER-2 status and absence of EGFR-activating mutations. Mod Pathol; 8:1027–33.
13. Chakravarti, A., Winter, K., Wu, C. L., Kaufman, D., Hammond, E., Parliament, M.,

Tester, W., Hagan, M., Grignon, D., Heney, N., Pollack, A., Sandler, H., Shipley, W. (2005) Expression of the epidermal growth factor receptor and Her-2 are predictors of favorable outcome and reduced complete response rates, respectively, in patients with muscle-invading bladder cancers treated by concurrent radiation and cisplatin-based chemotherapy: a report from the Radiation Therapy Oncology Group. Int J Radiat Oncol Biol Phys; 2:309–17.

14. Ding, Y., Wang, G., Ling, M. T., Wong, Y. C., Li, X., Na, Y., Zhang, X., Chua, C. W., Wang, X., Xin, D. (2006) Significance of Id-1 up-regulation and its association with EGFR in bladder cancer cell invasion. Int J Oncol; 4:847–54.
15. Lassus, H., Sihto, H., Leminen, A., Joensuu, H., Isola, J., Nupponen, N., Butzow, R. (2006) Gene amplification, mutation, and protein expression of EGFR and mutations of ERBB2 in serous ovarian carcinoma. J Mol Med; 8:671–81.
16. Stadlmann, S., Gueth, U., Reiser, U., Diener, P.-A., Zeimet, A. G., Wight, E., Mirlacher, M., Sauter, G., Mihatsch, M. J., Singer, G. (2006) Epithelial growth factor receptor status in primary and recurrent ovarian cancer. Mod Pathol; 4:607–10.
17. Doger, F. K., Meteoglu, I., Tuncyurek, P., Okyay, P., Cevikel, H. (2006) Does the EGFR and VEGF expression predict the prognosis in colon cancer? Eur Surg Res; 6:540–4.
18. Rohit, B., Beiyun, C., David, S. K., Leonard, B. S., Cyrus, H., Laura, H. T., William, G., Julie, T.-F., Philip, B. P., Jing, Q., Jinru, S. (2006) Comparison of two antibodies for immunohistochemical evaluation of epidermal growth factor receptor expression in colorectal carcinomas, adenomas, and normal mucosa. Cancer; 8:1857–62.
19. Jacobs, T. W., Prioleau, J. E., Stillman, I. E., Schnitt, S. J. (1996) Loss of tumor marker-immunostaining intensity on stored paraffin slides of breast cancer. J Natl Cancer Inst; 15:1054–9.
20. Manne, U., Myers, R. B., Srivastava, S., Grizzle, W. E. (1997) Re: loss of tumor marker-immunostaining intensity on stored paraffin slides of breast cancer. J Natl Cancer Inst; 8:585–6.
21. Bertheau, P., Cazals-Hatem, D., Meignin, V., de Roquancourt, A., Verola, O., Lesourd, A., Sene, C., Brocheriou, C., Janin, A. (1998) Variability of immunohistochemical reactivity on stored paraffin slides. J Clin Pathol; 5:370–4.
22. Mirlacher, M., Kasper, M., Storz, M., Knecht, Y., Durmuller, U., Simon, R., Mihatsch, M. J., Sauter, G. (2004) Influence of slide aging on results of translational research studies using immunohistochemistry. Mod Pathol; 11:1414–20.
23. Schlomm, T., Iwers, L., Kirstein, P., Jessen, B., Kollermann, J., Minner, S., Passow-Drolet, A., Mirlacher, M., Milde-Langosch, K., Graefen, M., Haese, A., Steuber, T., Simon, R., Huland, H., Sauter, G., Erbersdobler, A. (2008) Clinical significance of p53 alterations in surgically treated prostate cancers. Mod Pathol; 21(11):1371–8.
24. Darby, I. A., Bisucci, T., Desmouliere, A., Hewitson, T. D. (2006) In situ hybridization using cRNA probes: isotopic and nonisotopic detection methods. Methods Mol Biol; 326:17–31.
25. Barlund, M., Forozan, F., Kononen, J., Bubendorf, L., Chen, Y., Bittner, M. L., Torhorst, J., Haas, P., Bucher, C., Sauter, G., Kallioniemi, O. P., Kallioniemi, A. (2000) Detecting activation of ribosomal protein S6 kinase by complementary DNA and tissue microarray analysis. J Natl Cancer Inst; 15:1252–9.
26. Hanley, K. Z., Siddiqui, M. T., Lawson, D., Cohen, C., Nassar, A. (2009) Evaluation of new monoclonal antibodies in detection of estrogen receptor, progesterone receptor, and Her2 protein expression in breast carcinoma cell block sections using conventional microscopy and quantitative image analysis. Diagn Cytopathol; 4:251–7.
27. Moeder, C. B., Giltnane, J. M., Moulis, S. P., Rimm, D. L. (2009) Quantitative, fluorescence-based in-situ assessment of protein expression. Methods Mol Biol; 520:163–75.

Chapter 3

Representativity of TMA Studies

Guido Sauter

Abstract

The smaller the portion of a tumor sample that is analyzed becomes, the higher is the risk of missing important histological or molecular features that might be present only in a subset of tumor cells. Many researchers have, therefore, suggested using larger tissue cores or multiple cores from the same donor tissue to enhance the representativity of TMA studies. However, numerous studies comparing the results of TMA studies with the findings from conventional large sections have shown that all well-established associations between molecular markers and tumor phenotype or patient prognosis can be reproduced with TMAs even if only one single 0.6 mm tissue spot is analyzed. Moreover, the TMA technology has proven to be superior to large section analysis in finding new clinically relevant associations. The high number of samples that are typically included in TMA studies, and the unprecedented degree of standardization during TMA experiments and analysis often give TMA studies an edge over traditional large-section studies.

Key words: Tissue microarray, TMA representativity, TMA core diameter, Tissue heterogeneity, Translational research

1. Introduction

When the tissue microarray (TMA) technology was first published in 1998 (1), many researchers were concerned about the small size of the tissue spots. Tissues are typically heterogeneous, with areas of different histological or even genetic features in case of tumor tissues. Significant heterogeneity may occur between different areas of a tumor. For example, genetic differences have been found between different cells of the same tumor (2), between invasive and non-invasive parts (3–5), between primary tumors and metastases (6), or between multifocal tumors from the same organ (7). These findings can be explained with genetic instability resulting in a parallel development of multiple different tumor cell subclones with partly different biological and genetic properties.

Ronald Simon (ed.), *Tissue Microarrays: Methods and Protocols*, Methods in Molecular Biology, vol. 664,
DOI 10.1007/978-1-60761-806-5_3, © Springer Science+Business Media, LLC 2010

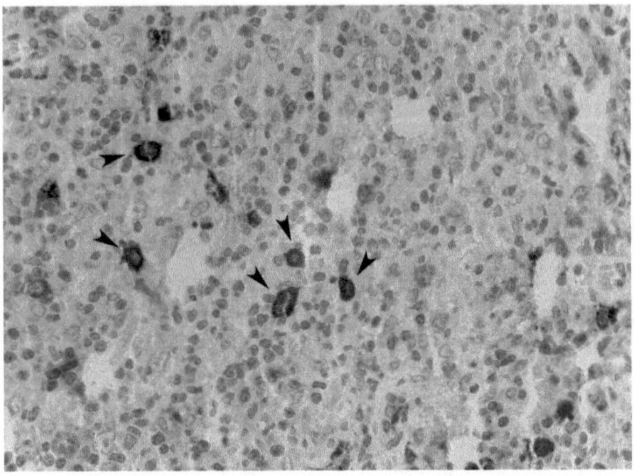

Fig. 1. Hodgkin's lymphoma as an example for tumor heterogeneity. The neoplastic Reed Sternberg cells (*arrowheads*) are marked by immunostaining detecting Epstein–Barr virus latent membrane antigen. Note that more than 90% of the cells are normal lymphocytes.

In some tumors, these subclones may be well separated from each other, while others may show a heterogeneous mixture of tumor cells. Such intratumoral heterogeneity is often found for the proliferation marker Ki-67, the expression of which can be highly variable in different areas of a tumor. Obviously, tissue heterogeneity is the most critical issue in tissue analysis, and so it was no surprise that many researchers initially disbelieved that TMAs with spots measuring only 0.6 mm in diameter would be representative for the whole tumor. For example, Hodgkin's lymphoma predominantly consists of reactive inflammatory cells with only few dispersed neoplastic Hodgkin's or Reed Sternberg cells (Fig. 1). Such tumors obviously bear inherent challenges for analyses on TMAs.

2. The Theory of Representativity

It is important to note that all of these concerns are based on the assumption that conventional whole sections – the "gold standard" for molecular tumor tissue analysis – are representative of the entire tumor bulk. However, considering the tiny volume of 0.0018 cm^3 of a typical whole tissue section (3×2 cm, thickness 3 μm) in comparison to a large tumor bulk measuring 5 cm in diameter (65 cm^3), it becomes evident that this assumption is unlikely to be true. In this example, the whole tissue section covers only 1/36,000 of the tumor, while a 0.6 mm tissue spot with a volume of 0.00000085 cm^3 is about 1/2,000 of the large section. These figures not only demonstrate that the "representativity problem" already exists between the whole tumor and the conventional large section, but also indicate that it is a magnitude (in the example, about 18 times) higher between the whole tumor and the "gold standard" large section than between the large section and the TMA spot (Fig. 1). All the studies comparing TMA results based on one, two, three or more punches from one donor block with the results obtained from large sections of the corresponding donor block therefore do not answer the most relevant question (8–12): To what extent is the TMA technique and/or the large section analysis suited to identify clinically relevant molecular markers?

3. Large Sections Analysis Versus TMA Analysis

In order to systematically compare the suitability of conventional large sections or TMAs to find clinically relevant associations between molecular features and clinico-pathological data, it is necessary to compare large section and TMA data with clinical outcome information. Few such studies exist, obviously because of the enormous workload involved. In one study, we showed that histologic grading and Ki-67 labeling index analysis on TMAs was equally relevant for predicting prognosis in bladder cancer as if the analysis was performed on large section (13). For another one of these studies, we manufactured a TMA consisting of three cores from the tumor periphery and of one core from the tumor center of 553 breast cancers and analyzed corresponding large sections by immunohistochemistry for ER, PR, and p53 expression (8). When the results of the TMA analysis were compared to the results of the large section analysis, highly concordant findings were seen for ER and PR. More than 80% of the ER- or PR-positive tumors were identified even if only one of the four replicate TMA was analyzed, and almost all (>98%) positive tumors were found if the results of the four arrays were combined. However, the TMA analysis yielded discrepant results for p53. Between 15 and 21% of p53-positive tumors were found in the four individual TMAs, and a combination of the results (i.e., a tumor is considered positive if at least one of the four TMAs yielded positive staining) resulted in only 24% p53-positive cases. In contrast, 43% p53-positive tumors were found when the conventional large sections were analyzed. All TMA sections and all large sections had been read by the same experienced gynecopathologist (Joachim Torhorst) using exactly identical criteria. If our study only involved a comparison of TMA versus large section results (as most studies analyzing TMA representativity were designed for), our findings had suggested that about half of the p53 positive tumors had been missed in the TMA analysis even though multiple punches per tumor had been analyzed. The comparison with follow-up data showed, however, that the TMA results were "better" than the large section data. The expected prognostic difference between p53-positive and p53-negative tumors was found in each of the four TMA analyses, but not in the large section analysis (Fig. 2). A reanalysis of the combined data from the large sections and the TMAs showed that 111 tumors, which had been rated p53 positive in the large section analysis but were found p53 negative in the TMAs, had a similarly favorable prognosis as 268 tumors with a concordantly p53-negative result in both the large section and TMA analysis (Fig. 2). Obviously, in the large section analysis, either focal staining artifacts had been overinterpreted as being p53 positive, or small areas of truly p53-positive cells, which had not been hit in the TMA spots, had no biological significance for the clinical course of the tumor. In any case, this example shows that TMAs can be even superior to large sections to detect clinically relevant associations between molecular findings and tumor phenotype.

We conclude from these data that TMA studies have some advantages over large section analyses, which may outweigh the disadvantage of limited tissue availability. These advantages include

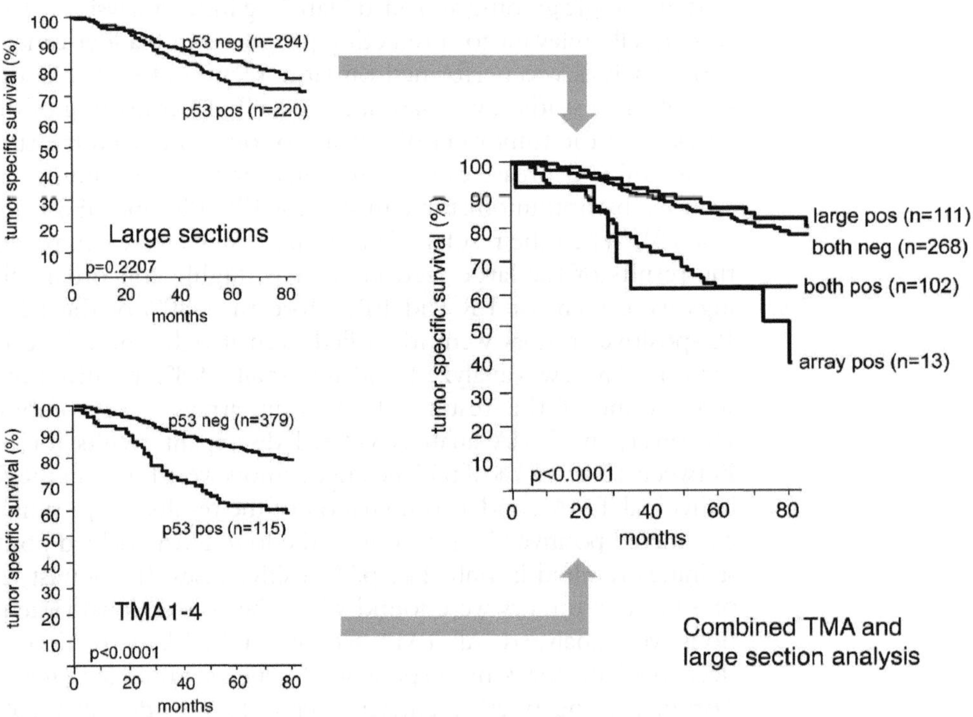

Fig. 2. Analysis of p53 immunostaining in >500 breast cancers analyzed on conventional large sections and in a TMA format. The known prognostic relevance was found in the TMA analysis only. A combined reanalysis revealed that the 111 cancers that had a positive IHC result in the large section analysis ("*large pos*") but were negative in the TMA analysis had no adverse impact on patient survival.

an unprecedented standardization in both staining and analysis of the staining. In large sections, even an expert pathologist, has to "select" the best representative area for interpretation. On a TMA, the interpretation of the staining is limited to the staining intensity. Obviously it is easier to consistently interpret staining intensities, if all cases of a study are located next to each other on one section. The representativity of TMAs is also demonstrated by the fact, that virtually all TMA studies analyzing large enough numbers of tumors, were able to fully reproduce previously well-established associations between molecular changes and clinico-pathological parameters (8, 11, 13–15). Likewise, Ruiz et al. successfully reproduced the known prognostic relevance of the highly heterogeneous proliferation marker Ki-67 expression in a large TMA study including more than 1,900 breast cancers (11). This is even true for studies analyzing difficult tumor types like Hodgkin's lymphoma (HL) (16–18). At least two studies have confirmed that HL-TMAs are reliable and representative. Hedvat et al. (17) analyzed an HL-TMA for CD20, CD30, CD15, Epstein–Barr virus (EBV) latent membrane protein 1 (LMP-1), EBER-1, and EBER-2 expression, and Tzankov et al. (18) also found 100% concordant LMP-1 expression in Hodgkin's and Reed-Sternberg (HRS) on conventional full sections with the corresponding tumor cores on the TMA.

4. How Many Cores?

It is our opinion that one 0.6-mm core per tumor is the ideal procedure for TMA manufacturing. This opinion is supported by a large number of TMA studies that confirm the known prognostic relevance of almost all previously established clinically useful biomarkers, for instance, between HER2 alterations and survival in breast cancer patients (15), between vimentin expression and prognosis in kidney cancer (14), as well as between the Ki67 labeling index and prognosis in urinary bladder cancer (13), and breast cancer (11).

Arguments against using two or more spots per tumor include additional and unnecessary work as well as statistical bias. A statistical bias is introduced if the amount of tissue (number of spots) varies between the tumors of one study. All studies analyzing the impact of multiple samples per cancer have shown that the fraction of positive cases increases with the number of analyzed spots (9, 12, 19–24). Obviously, the overall likelihood of a positive result is higher in cancers with more interpretable spots than in cancers with only one or a few interpretable spots. To avoid such a bias, the analysis might be limited to samples with identical numbers of interpretable spots (or to one spot per tumor only), but if all analyzable spots are included into the statistical analysis, special rules have to be defined on how to handle missing data from noninterpretable replicate spots or discrepant findings between replicate spots, making the analysis much more complicated.

Prostate cancer is a frequent tumor considered to be particularly heterogeneous. Several studies have analyzed whether the analysis of multiple cores has advantages in tumor TMA studies, including prostate cancer. Most studies indeed concluded that multiple samples, for example 3–4, offer advantages over our one core per tumor strategy (21, 25). However, the majority of them did not compare their data with clinical outcome. On the basis of our discussions above, we do not consider studies comparing large section data with TMA data meaningful in the absence of clinical parameters.

Only one study compared their results with outcome data. In this project, Rubin et al. (21) analyzed the Ki-67-labeling index (LI; fraction of cell nuclei positive for staining with the monoclonal antibody Ki-67) in ten separate cores of 88 prostate cancers. In this study, three to four cores were found to give the best association with PSA recurrence in these 88 patients. The authors concluded that 3–4 cores are required for optimal prostate cancer studies. However, the use of sufficiently large patient cohorts enables the analysis of molecular features with one 0.6 mm spot TMAs also in prostate cancer. For example, we found significant associations between expression of Egfr (26), PSA (27), p53 (28), CD10 (29), or reduced mast cell density (30) and adverse features of prostate cancer, including high Gleason grade, advanced stage, and tumor recurrence in our TMA containing >2,000 prostate cancers with follow-up data. Others have found significant associations between Egfr, EpCAM, Her2, HSP 90, ILK, Ki-67, MMP-2, MUC-1, p53, Syndecan-1 and gleason grade (31), HIF-1 alpha, VEGF, Osteopontin and time to biochemical failure (32), and WDR19 and time to biochemical failure (33) in TMAs from prostate cancers containing only one single tissue spot per tumor.

There are only two examples where we use more than one spot per tumor in our own studies. This applies to studies for which our patient sets are so small that losing a few cases due to insufficient tissue on a TMA makes a difference. For example, if we had a set of 88 prostate cancers with prognosis data, we would also add multiple cores on our TMA as suggested by Rubin et al. (21). The second example is comparative studies, for instance, between primary tumors and matched metastases (34). In order to have as many interpretable pairs as possible, it is advisable to add two cores per tumor on a TMA. Otherwise, a pair is always lost if either the primary tumor or the metastasis is lost due to insufficient tumor on the spot.

5. What Is the Best Core Diameter?

There is no general rule for the "optimal" core diameter. Rather, the size of a TMA spot should be adjusted according to the type of arrayed tissue.

In cancer TMAs, we strongly feel that small spots (e.g., 0.6 mm) have important advantages over large spots (e.g., 1 mm and above). As discussed earlier, the difference between a small and a large tissue spot or a conventional large section is virtually negligible with respect to the size of a large tumor bulk. However, small tissue spots often contain either only tumor cells, or no tumor cells at all. This leads to a most standardized analysis per tissue sample. The larger the arrayed spot, the higher the likelihood to include non-neoplastic cells and to increase the variability of the amount of cancer tissue analyzed per patient. In theory, smaller spots (e.g., 0.1 mm) would be even superior over 0.6 mm spots, because more cores can be placed in one TMA block, and the analysis becomes easier and faster because only small amounts of homogeneous tissues must be analyzed per tumor. However, core diameters below 0.6 mm are technically difficult to manufacture because the tiny needles often bend or break during punching. Again, the best argument for using small tissue cores is that all known relevant associations between molecular markers and cancer phenotype as well as patient prognosis have been successfully reproduced using 0.6 mm TMA cores (8, 11, 13, 26–28).

For TMAs from normal tissue, it is often advisable to use larger tissue cores (e.g., 1.0–1.2 mm). Normal tissues should include all relevant cell types of the respective organ, and these might not be fully represented in 0.6 mm tissue spots.

In one study on atherosclerosis, we have even used up to 4 mm cores. This TMA was designed to analyze inflammatory infiltrates and the microvascular network in the arterial wall of iliac, carotid, and renal arteries. In this study, the core diameter was adjusted to include full-thickness arterial wall sectors (35).

References

1. Kononen, J., Bubendorf, L., Kallioniemi, A., Barlund, M., Schraml, P., Leighton, S., Torhorst, J., Mihatsch, M. J., Sauter, G., Kallioniemi, O. P. (1998) Tissue microarrays for high-throughput molecular profiling of tumor specimens. Nat Med; 7:844–7.
2. Khalique, L., Ayhan, A., Weale, M. E., Jacobs, I. J., Ramus, S. J., Gayther, S. A. (2007) Genetic intra-tumour heterogeneity in epithelial ovarian cancer and its implications for molecular diagnosis of tumours. J Pathol; 3:286–95.
3. Simon, R., Burger, H., Brinkschmidt, C., Bocker, W., Hertle, L., Terpe, H. J. (1998) Chromosomal aberrations associated with invasion in papillary superficial bladder cancer. J Pathol; 4:345–51.
4. Sauter, G., Mihatsch, M. J. (1998) Pussycats and baby tigers: non-invasive (pTa) and minimally invasive (pT1) bladder carcinomas are not the same! J Pathol; 4:339–41.
5. Robanus-Maandag, E. C., Bosch, C. A., Kristel, P. M., Hart, A. A., Faneyte, I. F., Nederlof, P. M., Peterse, J. L., van de Vijver,

M. J. (2003) Association of C-MYC amplification with progression from the in situ to the invasive stage in C-MYC-amplified breast carcinomas. J Pathol; 1:75–82.

6. Cho, E. Y., Han, J. J., Choi, Y. L., Kim, K. M., Oh, Y. L. (2008) Comparison of Her-2, EGFR and cyclin D1 in primary breast cancer and paired metastatic lymph nodes: an immunohistochemical and chromogenic in situ hybridization study. J Korean Med Sci; 6:1053–61.
7. Simon, R., Eltze, E., Schafer, K. L., Burger, H., Semjonow, A., Hertle, L., Dockhorn-Dworniczak, B., Terpe, H. J., Bocker, W. (2001) Cytogenetic analysis of multifocal bladder cancer supports a monoclonal origin and intraepithelial spread of tumor cells. Cancer Res; 1:355–62.
8. Torhorst, J., Bucher, C., Kononen, J., Haas, P., Zuber, M., Kochli, O. R., Mross, F., Dieterich, H., Moch, H., Mihatsch, M., Kallioniemi, O. P., Sauter, G. (2001) Tissue microarrays for rapid linking of molecular changes to clinical endpoints. Am J Pathol; 6:2249–56.
9. Camp, R. L., Charette, L. A., Rimm, D. L. (2000) Validation of tissue microarray technology in breast carcinoma. Lab Invest; 12:1943–9.
10. Hendriks, Y., Franken, P., Dierssen, J. W., De Leeuw, W., Wijnen, J., Dreef, E., Tops, C., Breuning, M., Brocker-Vriends, A., Vasen, H., Fodde, R., Morreau, H. (2003) Conventional and tissue microarray immunohistochemical expression analysis of mismatch repair in hereditary colorectal tumors. Am J Pathol; 2:469–77.
11. Ruiz, C., Seibt, S., Al Kuraya, K., Siraj, A. K., Mirlacher, M., Schraml, P., Maurer, R., Spichtin, H., Torhorst, J., Popovska, S., Simon, R., Sauter, G. (2006) Tissue microarrays for comparing molecular features with proliferation activity in breast cancer. Int J Cancer; 9:2190–4.
12. Hoos, A., Urist, M. J., Stojadinovic, A., Mastorides, S., Dudas, M. E., Leung, D. H., Kuo, D., Brennan, M. F., Lewis, J. J., Cordon-Cardo, C. (2001) Validation of tissue microarrays for immunohistochemical profiling of cancer specimens using the example of human fibroblastic tumors. Am J Pathol; 4:1245–51.
13. Nocito, A., Bubendorf, L., Tinner, E. M., Suess, K., Wagner, U., Forster, T., Kononen, J., Fijan, A., Bruderer, J., Schmid, U., Ackermann, D., Maurer, R., Alund, G., Knonagel, H., Rist, M., Anabitarte, M., Hering, F., Hardmeier, T., Schoenenberger, A. J., Flury, R., Jager, P., Fehr, J. L., Schraml, P., Moch, H., Mihatsch, M. J., Gasser, T., Sauter, G. (2001) Microarrays of bladder cancer tissue are highly representative of proliferation index and histological grade. J Pathol; 3:349–57.
14. Moch, H., Schraml, P., Bubendorf, L., Mirlacher, M., Kononen, J., Gasser, T., Mihatsch, M. J., Kallioniemi, O. P., Sauter, G. (1999) High-throughput tissue microarray analysis to evaluate genes uncovered by cDNA microarray screening in renal cell carcinoma. Am J Pathol; 4:981–6.
15. Barlund, M., Forozan, F., Kononen, J., Bubendorf, L., Chen, Y., Bittner, M. L., Torhorst, J., Haas, P., Bucher, C., Sauter, G., Kallioniemi, O. P., Kallioniemi, A. (2000) Detecting activation of ribosomal protein S6 kinase by complementary DNA and tissue microarray analysis. J Natl Cancer Inst; 15:1252–9.
16. Garcia, J. F., Camacho, F. I., Morente, M., Fraga, M., Montalban, C., Alavaro, T., Bellas, C., Castano, A., Diez, A., Flores, T., Martin, C., Martinez, M. A., Mazorra, F., Menarguez, J., Mestre, M. J., Mollejo, M., Saez, A. I., Sanchez, L., Piris, M. A. (2002) Hodgkin's and Reed-Sternberg cells harbor alterations in the major tumor suppressor pathways and cell-cycle checkpoints: analyses using tissue-microarrays. Blood; 101:681–89.
17. Hedvat, C. V., Hegde, A., Chaganti, R. S., Chen, B., Qin, J., Filippa, D. A., Nimer, S. D., Teruya-Feldstein, J. (2002) Application of tissue microarray technology to the study of non-Hodgkin's and Hodgkin's lymphoma. Hum Pathol; 10:968–74.
18. Tzankov, A., Zimpfer, A., Lugli, A., Krugmann, J., Went, P., Schraml, P., Maurer, R., Ascani, S., Pileri, S., Geley, S., Dirnhofer, S. (2003) High-throughput tissue microarray analysis of G1-cyclin alterations in classical Hodgkin's lymphoma indicates overexpression of cyclin E1. J Pathol; 2:201–7.
19. Hoos, A., Stojadinovic, A., Mastorides, S., Urist, M. J., Polsky, D., Di Como, C. J., Brennan, M. F., Cordon-Cardo, C. (2001) High Ki-67 proliferative index predicts disease specific survival in patients with high-risk soft tissue sarcomas. Cancer; 4:869–74.
20. Hoos, A., Stojadinovic, A., Singh, B., Dudas, M. E., Leung, D. H., Shaha, A. R., Shah, J. P., Brennan, M. F., Cordon-Cardo, C., Ghossein, R. (2002) Clinical significance of molecular expression profiles of Hurthle cell tumors of the thyroid gland analyzed via tissue microarrays. Am J Pathol; 1:175–83.
21. Rubin, M. A., Dunn, R., Strawderman, M., Pienta, K. J. (2002) Tissue microarray sampling

strategy for prostate cancer biomarker analysis. Am J Surg Pathol; 3:312–9.

22. Engellau, J., Akerman, M., Anderson, H., Domanski, H. A., Rambech, E., Alvegard, T. A., Nilbert, M. (2001) Tissue microarray technique in soft tissue sarcoma: immunohistochemical Ki-67 expression in malignant fibrous histiocytoma. Appl Immunohistochem Mol Morphol; 4:358–63.
23. Fernebro, E., Dictor, M., Bendahl, P. O., Ferno, M., Nilbert, M. (2002) Evaluation of the tissue microarray technique for immunohistochemical analysis in rectal cancer. Arch Pathol Lab Med; 6:702–5.
24. Zhang, D., Salto-Tellez, M., Putti, T. C., Do, E., Koay, E. S. (2003) Reliability of tissue microarrays in detecting protein expression and gene amplification in breast cancer. Mod Pathol; 1:79–84.
25. Kristiansen, G., Fritzsche, F. R., Wassermann, K., Jager, C., Tolls, A., Lein, M., Stephan, C., Jung, K., Pilarsky, C., Dietel, M., Moch, H. (2008) GOLPH2 protein expression as a novel tissue biomarker for prostate cancer: implications for tissue-based diagnostics. Br J Cancer; 6:939–48.
26. Schlomm, T., Kirstein, P., Lwers, L., Daniel, B., Steuber, T., Walz, J., Chun, F. H. K., Haese, A., Kollermann, J., Graefen, M., Huland, H., Sauter, G., Simon, R., Erbersdobler, A. (2007) Clinical significance of epidermal growth factor receptor protein overexpression and gene copy number gains in prostate cancer. Clin Cancer Res; 22:6579–84.
27. Erbersdobler, A., Isbarn, H., Steiner, I., Schlomm, T., Chun, F., Mirlacher, M., Sauter, G. (2009) Predictive value of prostate-specific antigen expression in prostate cancer: a tissue microarray study. Urology; 74:1169–73.
28. Schlomm, T., Iwers, L., Kirstein, P., Jessen, B., Kollermann, J., Minner, S., Passow-Drolet, A., Mirlacher, M., Milde-Langosch, K., Graefen, M., Haese, A., Steuber, T., Simon, R., Huland, H., Sauter, G., Erbersdobler, A. (2008) Clinical significance of p53 alterations in surgically treated prostate cancers. Mod Pathol; 21:1371–8.
29. Fleischmann, A., Schlomm, T., Huland, H., Kollermann, J., Simon, P., Mirlacher, M., Salomon, G., Chun, F. H., Steuber, T., Simon, R., Sauter, G., Graefen, M., Erbersdobler, A. (2008) Distinct subcellular expression patterns of neutral endopeptidase (CD10) in prostate cancer predict diverging clinical courses in surgically treated patients. Clin Cancer Res; 23:7838–42.
30. Fleischmann, A., Schlomm, T., Kollermann, J., Sekulic, N., Huland, H., Mirlacher, M., Sauter, G., Simon, R., Erbersdobler, A. (2009) Immunological microenvironment in prostate cancer: High mast cell densities are associated with favorable tumor characteristics and good prognosis. Prostate; 69:976–81.
31. Zellweger, T., Ninck, C., Bloch, M., Mirlacher, M., Koivisto, P. A., Helin, H. J., Mihatsch, M. J., Gasser, T. C., Bubendorf, L. (2005) Expression patterns of potential therapeutic targets in prostate cancer. Int J Cancer; 4:619–28.
32. Vergis, R., Corbishley, C. M., Norman, A. R., Bartlett, J., Jhavar, S., Borre, M., Heeboll, S., Horwich, A., Huddart, R., Khoo, V., Eeles, R., Cooper, C., Sydes, M., Dearnaley, D., Parker, C. (2008) Intrinsic markers of tumour hypoxia and angiogenesis in localised prostate cancer and outcome of radical treatment: a retrospective analysis of two randomised radiotherapy trials and one surgical cohort study. Lancet Oncol; 4:342–51.
33. Lin, B., Utleg, A. G., Gravdal, K., White, J. T., Halvorsen, O. J., Lu, W., True, L. D., Vessella, R., Lange, P. H., Nelson, P. S., Hood, L., Kalland, K. H., Akslen, L. A. (2008) WDR19 expression is increased in prostate cancer compared with normal cells, but low-intensity expression in cancers is associated with shorter time to biochemical failures and local recurrence. Clin Cancer Res; 5:1397–406.
34. Simon, R., Nocito, A., Hubscher, T., Bucher, C., Torhorst, J., Schraml, P., Bubendorf, L., Mihatsch, M. M., Moch, H., Wilber, K., Schotzau, A., Kononen, J., Sauter, G. (2001) Patterns of her-2/neu amplification and overexpression in primary and metastatic breast cancer. J Natl Cancer Inst; 15:1141–6.
35. Fleiner, M., Kummer, M., Mirlacher, M., Sauter, G., Cathomas, G., Krapf, R., Biedermann, B. C. (2004) Arterial neovascularization and inflammation in vulnerable patients: early and late signs of symptomatic atherosclerosis. Circulation; 18:2843–50.

Chapter 4

Recipient Block TMA Technique

Martina Mirlacher and Ronald Simon

Abstract

New high-throughput screening technologies have led to the identification of hundreds of genes with a potential role in cancer or other diseases. One way to prioritize the leads obtained in such studies is to analyze a large number of tissues for candidate gene expression. The TMA methodology is now an established and frequently used tool for high-throughput tissue analysis. The recipient block technology is the "classical" method of TMA making. In this method, minute cylindrical tissue punches typically measuring 0.6 mm in diameter are removed from donor tissue blocks and are transferred into empty "recipient" paraffin blocks. Up to 1,000 different tissues can be analyzed in one TMA block. The equipment is affordable and easy to use in places where basic skills in histology are available.

Key words: Tissue microarray, High throughput tissue analysis, Recipient block technique

1. Introduction

In the current era of biochip technologies, the rate of discovery of new genes involved in cancer and other diseases has increased massively. Dozens of candidate genes are typically identified in a single DNA or protein chip experiment comparing expression patterns of diseased and normal tissues. The demand for validation studies of these new genes in diseased tissue specimens, especially human tumors, has grown at the same pace. To comprehensively study the potential clinical significance of promising candidate cancer genes, it is often necessary to analyze these genes in hundreds to thousands of well-characterized tumors with histopathological and clinical follow-up information. Such validation studies are virtually impossible to perform in a conventional slide-by-slide manner, not only because of the enormous workload but also because of the critical loss of precious tissue material connected with such studies. The tissue microarray (TMA) technology does significantly facilitate and accelerate tissue analyses

Ronald Simon (ed.), *Tissue Microarrays: Methods and Protocols*, Methods in Molecular Biology, vol. 664,
DOI 10.1007/978-1-60761-806-5_4, © Springer Science+Business Media, LLC 2010

by in-situ technologies (1). In this method, minute tissue cylinders are removed from hundreds of different primary tumor blocks and subsequently brought into one empty "recipient" paraffin block. Sections from such array blocks can then be used for simultaneous in situ analysis of hundreds to thousands of primary tumors on the DNA, RNA, and protein level. The TMA technique has a number of advantages as compared to the "sausage" block technique that has been introduced more than 20 years ago (2). The cylindrical shape and the small diameter of the specimen taken out of the donor block minimize the damage to the tissue and prevent exhaustion of the tissue blocks, while the original tissue blocks remain fully interpretable for all kinds of morphological and molecular analyses that may subsequently become necessary. Although there is no general rule to the "optimal" size of the tissue spots, a diameter of 0.6 mm has become a standard for tumor analysis. During the last decade, a multitude of TMA studies has shown that all known associations between molecular markers and clinical features of a tumor can be readily reproduced in tissue microarrays made from a single 0.6 mm core per donor tumor (3–8). These studies have also demonstrated that results obtained from TMA studies – despite the small core diameter – are representative for their donor tissues as long as sufficiently large numbers of tissue samples are included in the TMA. We strongly feel that TMA studies even have advantages over large section analyses, including an unprecedented standardization in both staining and analysis of the staining. In addition, the preselection of representative areas for coring by an expert pathologist during the TMA making process significantly facilitates and accelerates later TMA analysis. Larger core diameters (for instance 1.0–1.2 mm) are advisable for analysis of normal tissues, in order to make sure that all relevant cell types of the organ of interest are represented in a single tissue core. The recipient block technology is the "classical" method of TMA making, using an empty paraffin block with precored or predrilled holes as a mold for the TMA.

2. Materials

2.1. Sample Collection

1. Standard routine histology microscope for review of tissue sections.
2. Colored pens to mark representative areas on the slides, for example, red for tumor, blue for normal, and black for premalignant lesions.
3. Sufficient working space especially for large-scale projects that require extensive sorting of thousands of sections and blocks.

2.1.1. Preparing Recipient Blocks

1. PEEL-A-WAY Embedding Paraffin Pellets, melting point: 56–58°C (Polysciences Inc., PA, USA).
2. Slotted processing/embedding cassettes for routine histology, for example, EMS cat. # 70070 (Electron Microscopy Sciences Inc., PA, USA).
3. Stainless steel base molds for processing/embedding systems, for example, EMS cat. #62510-30 (Electron Microscopy Sciences Inc., PA, USA).
4. Filters/filter papers.
5. Oven for paraffin melting (70°C).

2.1.2. TMA Making

1. Tissue arrayer (currently there are multiple commercial vendors for tissue arrayers and supplies, including: http://www.beecher-instruments.com; http://www.millipore.com, http://www.biegler.com, http://www.veridiamtissuearrayers.com, http://www.pathologydevices.com).
2. Premanufactured empty paraffin-recipient blocks.
3. Illuminated magnifying lenses and supplies (e.g., Luxo U wave II/70, cat. #27950, Luxo Inc, Switzerland).

2.1.3. TMA Sectioning

1. Standard routine histology microtome and supplies (e.g., Leica SM2400, Leica Microsystems Inc., IL, USA).
2. Slide label printer (e.g., DAKO Seymour glass slide labeling system, product code S3416; DAKO A/S, Denmark) or special slide marker (e.g., Securline Marker II, Precision Dynamics Corporation, CA, USA).
3. Boxes for slide storage.
4. Refrigerator for slide storage.
5. Paraffin Sectioning Aid-System (Instrumedics Inc., NJ, USA; cat. # PSA) containing Ultraviolet Curing Lamp, Adhesive-Coated PSA Slides, TPC Solvent, TPC Solvent can, Hand roller, Tape windows.

3. Methods

3.1. TMA Manufacturing

3.1.1. Sample Collection

Although a device is needed to manufacture TMAs, it must be understood that most of the work (approximately 95%) is traditional pathology work that cannot be accelerated by improved (i.e., automated) tissue arrayers. This preparatory work is similar to what is needed for traditional studies involving "large" tissue sections. The major difference is the number of tissues involved, which can be an order of magnitude higher in TMA studies than in traditional projects. The different tasks related to sample collection are described below:

1. Define exactly the TMA to be made (see Note 1).
2. Generate a list of potentially suited tissues.
3. Collect all slides from these tumors from the archive.
4. One pathologist must review all sections from all candidate specimens to select the optimal slide (see Note 2).
5. Collect the tissue blocks that correspond to the selected slides.
6. These blocks and their corresponding marked slides must be matched and sorted in the order of appearance on the TMA.
7. Define the structure (outline) of the TMA and compose a file that contains the identification numbers of the tissues together with their locations and real coordinates (as they need to be selected on the arraying device). For the distance between the individual samples, 0.2 mm is recommended. To facilitate navigation on the TMA, we recommend arranging the tissues in multiple sections (e.g., quadrants). The distance between the quadrants may be 0.8 mm. In most laboratories capitalized letters define quadrants, whereas small letters and numbers define the coordinates within these quadrants. Examples of a TMA structure (outline) and data file containing the necessary information for making a TMA are given in Fig. 1 and Table 1 (see Note 3).

3.1.2. Preparing Recipient Blocks

In contrast to normal paraffin blocks, tissue microarray blocks are cut at room temperature. Therefore, a special type of paraffin is needed with a melting temperature between 55 and 58°C ("Peel-A-Way" paraffin, see Subheading 2). The paraffin is melted at 60°C, filtrated and poured in a stainless steel mold. A slotted plastic embedding cassette (as used in every histology lab) is then placed on the top of the warm paraffin. Carefully cool down the block to 4°C before removing it from the mold (see Note 4). Quality check of the recipient blocks is important because they must not contain air bubbles. Large recipient blocks (for example 30 × 45 × 10 mm) are easier to handle than the small blocks (for example 25–35 × 5 mm) that are typically used in routine histology labs.

3.1.3. TMA Making

Only if all this preparatory work has been done, a tissue-arraying device can be employed. Both manually operated and semiautomated systems are commercially available (see Subheading 2.1.3). Excellent TMAs can be produced in the hands of a talented and experienced person even with simple devices. However, optimal arrays can be expected only after a significant training period, mostly including several hundred, if not a few thousand punches. Patience, endurance, and keen eyesight are important prerequisites for operators of the manual tissue arrayers. Semiautomated

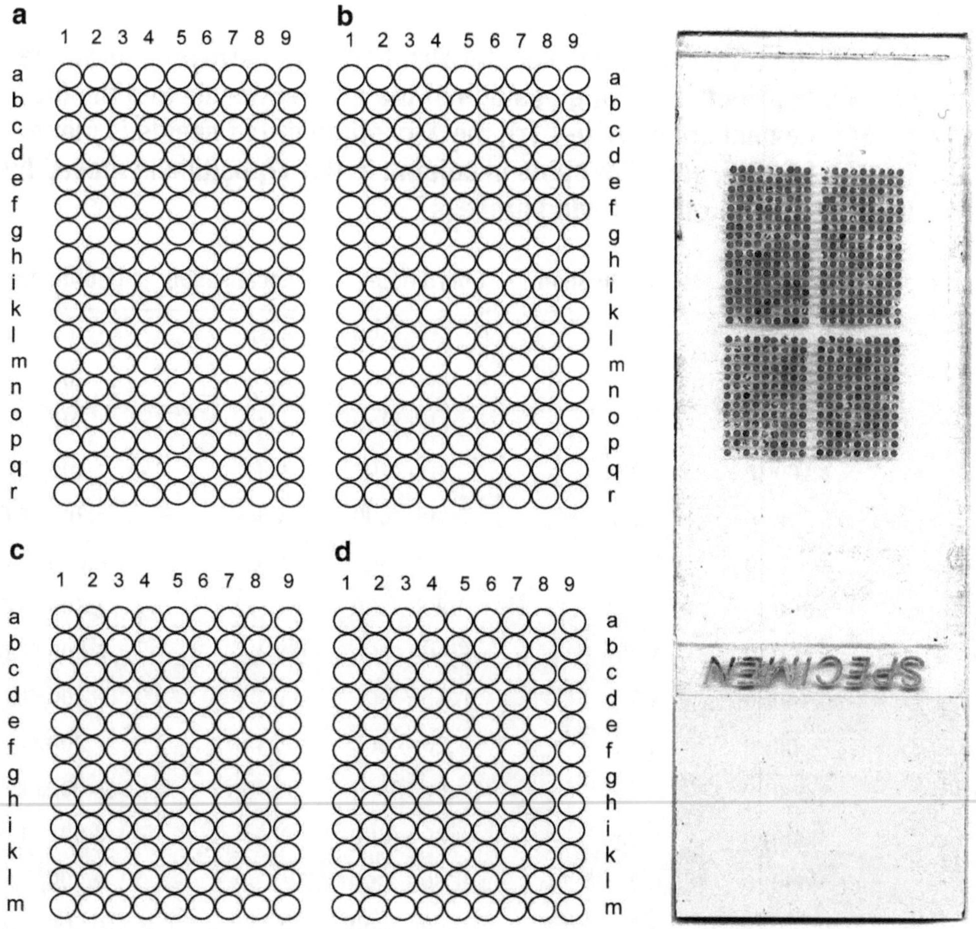

Fig. 1. TMA outline example. The TMA has been divided into four subsections to facilitate navigation during microscopy.

devices are available but these devices are very expensive and do not accelerate the TMA manufacturing process.

The TMA manufacturing process consists of four steps that are repeated for each sample placed on the TMA:

1. Punching a hole in an empty (recipient) paraffin block (see Note 5).
2. Removing and discarding the wax cylinder from the needle used for recipient block punching.
3. Removing a cylindrical sample from a donor paraffin block.
4. Placing the cylindrical tissue sample in the premade hole in the recipient block. Exact positioning of the tip of the tissue cylinder at the level of the recipient block surface is crucial for the quality and the yield of the TMA block. Releasing the tissue too deeply into the recipient block results in empty

Table 1
Template of an excel file linking the spot localization to the arrayer coordinate. The position 0/0 marks the starting position of the arrayer needle and the first spot (A1a). For the next spots (A1b–A1r), the *X*-position of the needle is moved by 800 μm (880/0, 1,600/0, etc.). For spots in column 2, the *Y*-position is moved by 800 μm before repeating the arraying process

Localisation	Coordinate	Localisation	Coordinate	Localisation	Coordinate
A 1a	0/0	A 2a	0/800	A 3a	0/1,600
A 1b	800/0	A 2b	800/800	A 3b	800/1,600
A 1c	1,600/0	A 2c	1,600/800	A 3c	1,600/1,600
A 1d	2,400/0	A 2d	2,400/800	A 3d	2,400/1,600
A 1e	3,200/0	A 2e	3,200/800	A 3e	3,200/1,600
A 1f	4,000/0	A 2f	4,000/800	A 3f	4,000/1,600
A 1g	4,800/0	A 2g	4,800/800	A 3g	4,800/1,600
A 1h	5,600/0	A 2h	5,600/800	A 3h	5,600/1,600
A 1i	6,400/0	A 2i	6,400/800	A 3i	6,400/1,600
A 1k	7,200/0	A 2k	7,200/800	A 3k	7,200/1,600
A 1l	8,000/0	A 2l	8,000/800	A 3l	8,000/1,600
A 1m	8,800/0	A 2m	8,800/800	A 3m	8,800/1,600
A 1n	9,600/0	A 2n	9,600/800	A 3n	9,600/1,600
A 1o	10,400/0	A 2o	10,400/800	A 3o	10,400/1,600
A 1p	11,200/0	A 2p	11,200/800	A 3p	11,200/1,600
A 1q	12,000/0	A 2q	12,000/800	A 3q	12,000/1,600
A 1r	12,800/0	A 2r	12,800/800	A 3r	12,800/1,600

spots in the first sections taken from the TMA block. Positioning the tissue cylinder too high causes empty spots in the last sections taken from this TMA. However, a too superficial location of the tissue cylinder is less problematic than a too deep position since protruding tissue elements can – to some extent – be leveled out after finishing the punching process. The use of a magnifying lens facilitates precise deposition of samples, especially for beginners. If all tissue elements are filled into the recipient block, the block is heated at 40°C for 10 min. Protruding tissue cylinders are then gently pressed deeper into the warmed TMA block using a glass slide.

3.1.4. Array Sectioning

Regular microtome sections may be taken from TMA blocks using standard microtomes. However, the more samples a TMA block contains, the more difficult regular cutting becomes. As a consequence, the number of slides of inadequate quality increases with the size of the TMA, and in turn, fewer sections from the TMA block can effectively be analyzed.

Using a tape sectioning kit (Instrumedics) facilitates cutting and leads to highly regular nondistorted sections (ideal for automated analysis). In addition, the tape system may prevent arrayed samples from floating off the slide, if very harsh pretreatment methods are used. However, the sticky glued slides have the disadvantage of increased background signals between the tissue spots in IHC analyses. The tissue samples themselves do not show increased non-specific background in IHC. The use of the tape sectioning system is described below:

1. An adhesive tape is placed on the TMA block in the microtome immediately before cutting.
2. A 3–5 μm section is cut. The tissue slice is now adhering to the tape.
3. The tissue slice is placed on a special "glued" slide (stretching of the tissue in a water bath or on a heating plate is not necessary).
4. The slide (tissue on the bottom) is then placed under UV light for 35 s. This leads to polymerization of the glue on the slide and on the tape.
5. Slides are placed into TPC solution (Instrumedics) at room temperature for 5–10 s. The tape can then be gently removed from the glass slide. The tissue remains on the slide.
6. Slides are dried at room temperature.

4. Notes

1. Make a comprehensive list of all tissues to be included into the TMA. Often TMA users realize that one critical control tissue has been forgotten only after completion of the TMA block. Include normal tissues of the organ of interest and – if possible – of a selection of other organs as well. The latter samples may be included in every TMA and be used as a standard control for the success of the IHC experiment.
2. If possible, tumors should be reclassified at that stage according to current classification schemes and tissue areas suited for subsequent punching should be marked. Different colors are recommended for marking different areas on one section

(for example, red for tumor, black for carcinoma in situ, blue for normal tissue). It is advisable to have a fresh HE-stained section if the actual block surface is not well reflected on the available stained section.

3. For unequivocal identification of individual samples on TMA slides, it is important to avoid a fully symmetrical TMA structure.
4. It is important not to cool down the paraffin on a cooling plate because of the risk of block damage. We cool down recipient paraffin blocks for 2 h at room temperature and then for 2 additional hours at 4°C. The blocks are then removed from the mold.
5. The hole may also be drilled into the paraffin block. Drilling is superior to punching because less force is applied to the paraffin block, making it more stable especially when large arrays are made. We have constructed our own arrayer where we have replaced the punch needle with a drill.

References

1. Kononen J, Bubendorf L, Kallioniemi A, Barlund M, Schraml P, Leighton S, Torhorst J, Mihatsch MJ, Sauter G, Kallioniemi OP. (1998) Tissue microarrays for high-throughput molecular profiling of tumor specimens. Nat Med 4:844–7.
2. Battifora H. (1986) The multitumor (sausage) tissue block: novel method for immunohistochemical antibody testing. Lab Invest 55:244–8.
3. Ruiz C, Seibt S, Al Kuraya K, Siraj AK, Mirlacher M, Schraml P, Maurer R, Spichtin H, Torhorst J, Popovska S, Simon R, Sauter G. (2006) Tissue microarrays for comparing molecular features with proliferation activity in breast cancer. Int J Cancer 118:2190–4.
4. Simon R, Nocito A, Hübscher T, Bucher C, Torhorst J, Schraml P, Bubendorf L, Mihatsch MJ, Moch H, Wilber K, Schötzau A, Kononen J, Sauter G. (2001) Patterns of her-2/neu amplification and overexpression in primary and metastatic breast cancer. J Natl Cancer Inst 93:1141–6.
5. Torhorst J, Bucher C, Kononen J, Haas P, Zuber M, Kochli OR, Mross F, Dieterich H, Moch H, Mihatsch M, Kallioniemi OP, Sauter G. (2001) Tissue microarrays for rapid linking of molecular changes to clinical endpoints. Am J Pathol 159:2249–56.
6. Schlomm T, Iwers L, Kirstein P, Jessen B, Kollermann J, Minner S, Passow-Drolet A, Mirlacher M, Milde-Langosch K, Graefen M, Haese A, Steuber T, Simon R, Huland H, Sauter G, Erbersdobler A. (2008) Clinical significance of p53 alterations in surgically treated prostate cancers. Mod Pathol 21:1371–8.
7. Bubendorf L, Kolmer M, Kononen J, Koivisto P, Mousses S, Chen Y, Mahlamaki E, Schraml P, Moch H, Willi N, Elkahloun AG, Pretlow TG, Gasser TC, Mihatsch MJ, Sauter G, Kallioniemi OP. (1999) Hormone therapy failure in human prostate cancer: analysis by complementary DNA and tissue microarrays. J Natl Cancer Inst 91:1758–64.
8. Nocito A, Bubendorf L, Tinner EM, Suess K, Wagner U, Forster T, Kononen J, Fijan A, Bruderer J, Schmid U, Ackermann D, Maurer R, Alund G, Knonagel H, Rist M, Anabitarte M, Hering F, Hardmeier T, Schoenenberger AJ, Flury R, Jager P, Fehr JL, Schraml P, Moch H, Mihatsch MJ, Gasser T, Sauter G. (2001) Microarrays of bladder cancer tissue are highly representative of proliferation index and histological grade. J Pathol 194:349–57.

Chapter 5

Protocol for Constructing Tissue Arrays by Cutting Edge Matrix Assembly

Thai Hong Tran, Justin Lin, Ashley Brooke Sjolund, Fransiscus Eri Utama, and Hallgeir Rui

Abstract

We present a protocol for construction of high-density tissue microarrays, *cutting edge matrix assembly*, which is based on repetitive sectioning and bonding of tissues. Maximized array density is achieved by a scaffold-free, self-supporting construction with rectangular array features that are bonded edge-to-edge, resulting in minimal wasted space between samples. Construction of the tissue array blocks from paraffin-embedded tissue involves initial bonding of primary tissue plates into multiple primary tissue stacks. This is achieved by taking a shaving of desired thickness from the face of each specimen block, trimming the shavings into a set of rectangular primary tissue plates, and bonding multiple plates into primary stacks of tissue. Each resulting primary tissue stack is then transversely cut to produce a set of secondary tissue plates that contains elements of each tissue represented in the primary stacks. Secondary plates from multiple primary sample stacks are then restacked and bonded into a secondary stack. The assembled secondary stack represents a laminate of laminates, which becomes the final array block. The final array block is then reembedded in paraffin and can be sectioned transversely using a microtome to yield micrometer thin sections that are transferred to glass slides for array display and analysis. This technology has facilitated the construction of arrays containing more than 10,000 tissue features on a standard glass slide.

Key words: Tissue microarrays, Cutting edge matrix assembly, Tissue analysis, Tissue array construction

1. Introduction

Tissue microarrays have become widely adopted for effective parallel in situ analysis of hundreds of tissues placed on single slides (1, 2). Traditionally, paraffin-embedded tissues are arrayed by transferring core punches into predrilled holes within a scaffold block of paraffin, and then sectioned transversely by a microtome to generate array sections. While core-based arraying continues to greatly advance tissue analyses, its limitations include restricted and inconsistent depth of cores, unpredictable tissue content and

Ronald Simon (ed.), *Tissue Microarrays: Methods and Protocols*, Methods in Molecular Biology, vol. 664,
DOI 10.1007/978-1-60761-806-5_5, © Springer Science+Business Media, LLC 2010

quality of the tissue cores beneath the surface of the tissue block, and limited sample numbers per slide because of excessive space taken up by the paraffin scaffold. In addition, feature sizes are restricted to a limited range of circular coring tools (e.g., 0.6, 1 and 2 mm). Overcoming these limitations, we have developed a complementary tissue arraying method that combines serial cutting and edge-to-edge bonding of samples to assemble a scaffold-free array matrix, cutting-edge matrix assembly (CEMA) (3, 4). Using CEMA, we have successfully placed more than 10,000 individual tissue sections onto a single histological glass slide. Here we provide a step-by-step guide to the construction of high-density CEMA tissue microarrays.

As discussed in greater detail in other chapters of this book, the central benefit of tissue microarrays is that parallel analyses of multiple tissues on the same slide allows for greater control of the multiple variables associated with all types of in situ analyses, including immunohistochemistry, in situ PCR, and in situ hybridization. For immunohistochemistry, each of the tissue features within the array is exposed to the same conditions during antigen retrieval, peroxidase block, antibody incubation, secondary antibodies or amplification steps, and enzyme reactions. These processes are difficult to keep consistent not only between different slides done in parallel, but also with slides that are processed on different days. Therefore, tissue microarrays are much more effective at revealing differences between macromolecular expression levels in multiple specimens than when individual slides of the same samples are analyzed one by one on individual slides and in different batches, even under the most stringent conditions. In addition, analysis of tissue microarray slides is highly cost-effective due to the low amounts of reagents needed per sample. Notably, once assembled, a tissue array can be sectioned and analyzed repeatedly for cost-effective, serial investigation of multiple parameters in the same samples.

CEMA-format arrays have been characterized as the next step in the evolution of the tissue microarray (5). We have successfully used CEMA tissue arrays to study hormone responses across large numbers of animal tissues (6) and for biomarker studies in human cancer specimens (7–9). CEMA tissue arrays are based on a self-supported construction that involves sequential bonding and transverse sectioning of tissue plate stacks. Array density is high due to the rectangular sample features and elimination of space loss due to a structural scaffold; combined, this allows for maximal use of available space and high packing density. In contrast, core-based tissue arrays require placement of cylindrical tissue cores into a paraffin scaffold, typically resulting in arrays where 50% or more of the array area represents paraffin. Furthermore, because CEMA uses surface shavings and not core punches for array construction, this leaves the original tissue block pristine

Table1
Comparison between CEMA and traditional core-based array technologies

Features of tissue microarray technologies	Core-based arrays	CEMA arrays
Cost-effective parallel in situ analysis of samples	Yes	Yes
Minimal slide-to-slide variation – standardization	Yes	Yes
Low reagent and tissue consumption	Yes	Yes
Variety of in situ analysis methods	Yes	Yes
Retains tissue architecture	Yes	Yes
Amenable to automated IHC scoring	Yes	Yes
Number of features possible per slide	**Hundreds**	**Thousands**
Known tissue quality and content along depth of array	Uncertain	Yes
Virtually unlimited scalability of array features	No	Yes
Unaffected quality of tissue block after sampling	No	Yes
Specialized equipment required	Yes	No

and without any reduction in quality associated with multiple punch holes. A comparative list of features of CEMA and core-based tissue arrays is provided in Table 1. Examples of CEMA tissue arrays are shown in Fig. 1.

The practical construction of CEMA arrays is straightforward and achievable with standard pathology laboratory equipment. Building of the array blocks involves a two-step procedure (Fig. 2). First, tissue shavings of desired thickness are generated and trimmed into primary plates of desired size (Fig. 2a). For instance, 1 cm by 0.5 cm works well. Primary tissue plates are then stacked and bonded using surgical glue into primary stacks or laminates (Fig. 2b). For instance, 100 primary tissue plates can be assembled into ten stacks of ten plates each (Fig. 2c). Second, each primary tissue stack is then transversely cut to produce a secondary tissue plate that contains elements of each tissue represented in the primary stack. The secondary plate from each of the primary tissue stacks are then stacked and bonded using surgical glue into a secondary stack or laminate. This secondary stack becomes the final CEMA array block, and contains elements of all of the samples (Fig. 2d). Importantly, repeated cutting of secondary plates from the primary stacks and their assembly into additional secondary stacks will yield multiple copies of the final array block (Fig. 2d). Following these two construction steps, the array block is finally sectioned transversely using a microtome to yield micrometer thin array sections for transfer to support slides for analysis (Fig. 2e).

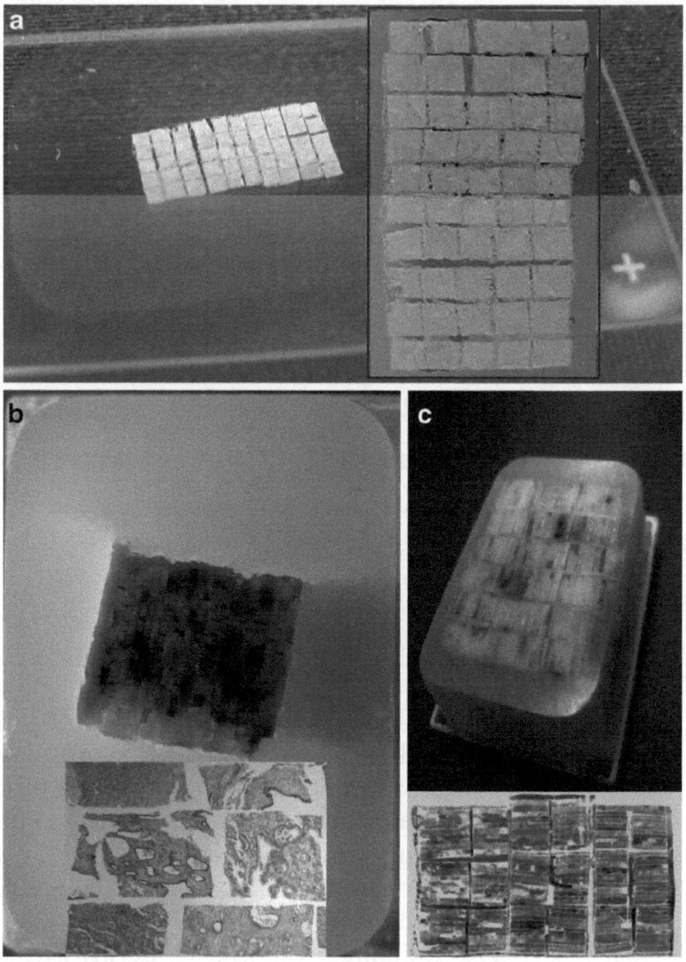

Fig. 1. Examples of CEMA array blocks. (**a**) Miniarray of 50 (5 × 10) liver tissue specimens represented by microtome section on glass slide. Insert shows magnified view of section before deparaffinization. (**b**) A multitumor CEMA tissue array of 200 tissue features (10 each of the 10 most common solid tumors at twofold redundancy). Insert represents H&E stained section. (**c**) A breast cancer progression array with 180 specimens, including normal, ductal carcinoma in situ, invasive cancer and metastases. Insert represents H&E stained section.

Once a CEMA array block has been constructed, 3–5 μm thin microtome sections allow production of a large number of array copies for serial analyses of a corresponding large number of parameters or analytes. An array block depth of 1 cm allows cutting of 2,000 or more usable array sections. Furthermore, the primary sample stacks yields 5–10 replicate secondary sample plates for assembly and construction of 5–10 replicate CEMA array blocks. Therefore as many as 10,000 parameters such as macromolecules can be investigated from the same set of tissues, opening new possibilities in array-based screening. During arraying of tumor tissues, it is especially difficult to predict the depth of quality tumor tissue when cores are punched from the surface

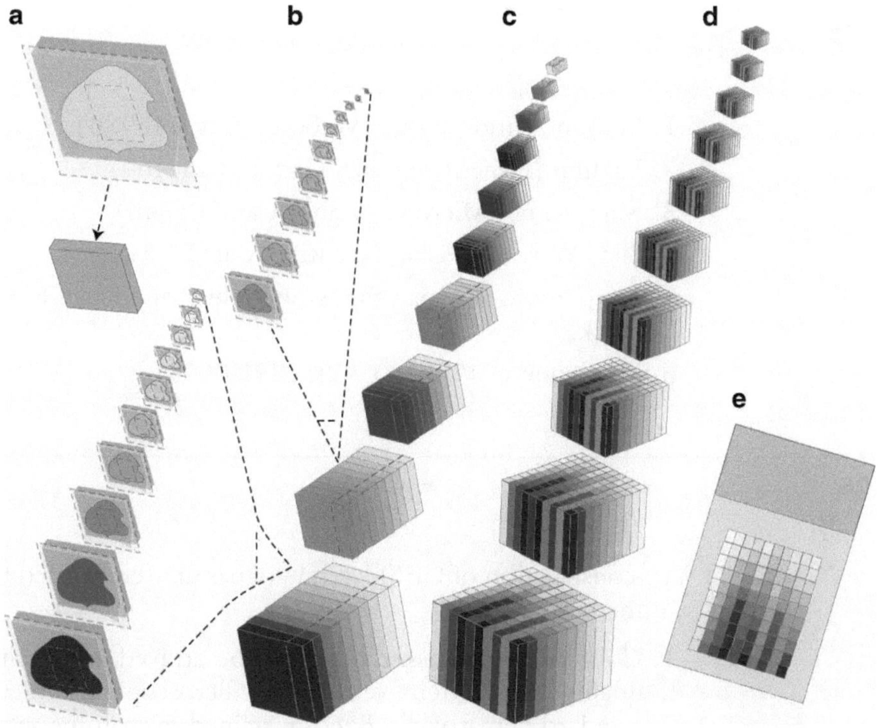

Fig. 2. Overview of CEMA tissue array construction. (**a**) Initial cutting of thick section (*primary plate*) from tissue block. (**b**) Bonding and stacking of primary tissue plates into primary tissue stacks or laminates. (**c**) Cutting of secondary plates and assembly to form CEMA array blocks (**d**). (**e**) Final microtome array section on glass slide.

of archival blocks. It is therefore not uncommon that 50% or more of the cores are effectively lost after 100–150 sections of conventional arrays (5). CEMA has the potential to yield higher quality tumor arrays, because the technology allows better verification of tumor content within the entire length of a thin superficial plate shaved off from the tumor block surface. In cases where the area of interest in a given sample (e.g., tumor) is smaller than the desired primary plate size, splicing of two or more plate fragments of the sample can be done to form a fused plate.

Another key benefit of CEMA arrays are the scalable array feature dimensions. The width of each feature corresponds to the thickness of the primary plates, and the length of each feature corresponds to the thickness of the secondary plates. When a microtome is used for cutting of plates, sample feature dimensions from as little as 5 μm up to several mm can be achieved, resulting in continuously tunable widths and lengths of array features. Assembly of small features into high-density microarrays is possible because microscopic sample elements are not handled individually, but instead collectively manipulated many at a time in sizeable continuous sheets. In contrast, core-based punching of tissue blocks is typically restricted to available hollow needles of diameter 0.6, 1, or 2 mm.

2. Materials

1. Vetbond surgical glue (Vetbond, Cat: 1469SB).
2. Paraffin (Fisher Scientific, Cat: 23-021-400).
3. Microtome (Microm, GmbH Cat: 905340).
4. Slide Warmer (Fisher Scientific, Cat: 12-594).
5. Deep mold for paraffin embedding of final CEMA array block.
6. Heat source (Vornado, Cat: EH1-0001).

3. Methods

The construction of CEMA arrays of paraffin-embedded tissues is outlined in Fig. 2.

1. Collect paraffin tissue blocks to be arrayed and verify tissue quality and content verified by microscopic examination of standard hematoxylin & eosin stained microtome sections.
2. Place paraffin tissue block in the holder of a standard laboratory microtome with the block surface parallel to the plane of the blade.
3. Trim block until the surface is even.
4. Advance paraffin block to the desired length (e.g., 50–2,000 μm), see Notes 1–2.
5. Gently heat the surface of tissue block with a hot air blower to soften the paraffin, see Note 3.
6. Cut thick section of paraffin block.
7. When cutting, place your thumb on tissue in block and press gently downward to prevent sliding of the tissue in the softened block caused by upward pressure by the blade. Move the blade up at a slow, even speed.
8. Place the newly cut thick section on a glass plate and while still soft, flatten it on the surface by sandwiching it with another glass plate and let it cool and harden.
9. Use a one-sided razor blade or other cutting tool to trim the tissue plate into the desirable size (e.g., 5 mm × 10 mm). The product is called a primary tissue plate.
10. Repeat this process to generate primary tissue plates from each of the samples that you want to array.
11. Place a small amount of surgical glue (e.g., Vetbond) on one side of tissue plate, see Note 4.

12. Build stacks of primary plates by pressing them together evenly and firmly after applying glue.
13. Reembed primary stacks in paraffin, so that tissue plates are perpendicular to the resulting block surface.
14. From primary stack blocks, make a transverse cut after gentle heating of the face of the block to generate a secondary plate at desired thickness, see Note 5.
15. Place the newly cut secondary plate on glass plate and flatten.
16. Bond secondary plates together using surgical glue (e.g., Vetbond).
17. Embed the resulting secondary stack in paraffin with each of the samples represented as a checkerboard on the surface of the resulting block.
18. Cut standard 3–5 μm thin sections and transfer array sections to glass slides, see Notes 6 and 7.
19. Deparaffinize and proceed with standard IHC or other labeling protocol, see Note 8.

4. Notes

1. The microtome needs to have a mechanism for horizontal advancement of the specimen without moving it up and down while advancing the block to the desired thickness of the thick section.
2. The thickness of the cut to generate primary plates will define the first dimension of the final rectangular array feature.
3. This will prevent cracking of the tissue during cutting and facilitate even movement of the knife.
4. Among usable cyanoacrylate glues, Vetbond is cheaper than Dermabond although both work fine.
5. The thickness of the cut from the primary stack to generate secondary plates will define the second dimension of the final rectangular array feature.
6. The scalable tissue features of CEMA arrays allow customized scaling of sample tissue area to be represented in the array. For instance, homogenous tissue set may be represented by smaller features than a heterogeneous. Furthermore, because CEMA arrays permit higher density packing of samples, more replicate samples can fit on a slide for increased redundancy when needed without compromising individual feature size.

7. Cyanoacrylate glue provides direct and lateral bonding of neighboring tissue elements to each other, resulting in CEMA array sections that behave as continuous sheets of fused sample material. Consequently, there is less chance of loss of individual features during sectioning and placement on slides, which can be a problem with core-based arrays where array samples are held together only by fragile paraffin. As a result, CEMA allowed effective transfer and adhesion of >10,000 features using regular flotation bath technique and Fisher ++ slides, without the use of tape-aid or highly adhesive Instrumedics slides.
8. Cyanoacrylate glue dissolves in xylene during deparaffinization.

Acknowledgments

This work was supported by Public Health Service grants R01-CA118740, R01-CA101841 (to H.R.), T32 CA09678 fellowship (to T.H.T.) from the National Institutes of Health (NIH) and grant PDF050415 from Susan G. Komen Foundation (to F.E.U.). Furthermore, this project is funded, in part, under a Commonwealth University Research Enhancement Program grant with the Pennsylvania Department of Health. The Department specifically disclaims responsibility for any analyses, interpretations, or conclusions. We thank Davina Miranda for expert help with graphic design and manuscript editing.

References

1. Kononen J, Bubendorf L, Kallioniemi A, et al. (1998) Tissue microarrays for high-throughput molecular profiling of tumor specimens [see comments]. Nat Methods 4(7):844–7.
2. Simon R, Sauter G. (2003) Tissue microarray (TMA) applications: implications for molecular medicine. Expert Rev Mol Med 5(26):1–12.
3. LeBaron MJ, Crismon HR, Utama FE, et al. (2005) Ultrahigh density microarrays of solid samples. Nat Methods 2(7):511–3.
4. Rui H, Lebaron MJ. (2005) Creating tissue microarrays by cutting-edge matrix assembly. Expert Rev Med Devices 2(6):673–80.
5. Rimm DL. (2005) Tissue microarrays without cores. Nat Methods 2(7):492–3.
6. LeBaron MJ, Ahonen TJ, Nevalainen MT, Rui H. (2007) In vivo response-based identification of direct hormone target cell populations using high-density tissue arrays. Endocrinology 148(3):989–1008.
7. Plotnikov A, Li Y, Tran TH, et al. (2008) Oncogene-mediated inhibition of glycogen synthase kinase 3 beta impairs degradation of prolactin receptor. Cancer Res 68(5):1354–61.
8. Lindsay J, Jiao X, Sakamaki T, et al. (2008) ErbB2 induces Notch1 activity and function in breast cancer cells. Clin Transl Sci 1(2):107–15.
9. Wang Y, Dean JL, Millar E, et al. (2008) Cyclin D1b is aberrantly regulated in response to therapeutic challenge and promotes resistance to estrogen antagonists. Cancer Res 68(14):5628–38.

Chapter 6

Hypodermic Needle Without Recipient Paraffin Block Technique

Andréa Rodrigues Cordovil Pires and Simone Rabello de Souza

Abstract

This technique allows building TMA blocks with more than 300 tissue cores without using a recipient paraffin block for the tissue cores and without using a commercial TMA builder instrument. It is based on the construction of TMA needles modifying conventional hypodermic needles to punch tissue cores from donor blocks. Tissue cores are punctured from donor blocks and attached by double-sided adhesive tape on a computer-generated paper grid used to align the cores on the block mold, which is filled with liquid paraffin. TMA blocks constructed using this method can be utilized as positive and negative controls for immunohistochemistry, histochemistry and other techniques, interlaboratory quality control and also in research, but never for diagnosis purposes. This technique has the following advantages: it is easy to reproduce, affordable, quick, and creates uniform blocks with more than 300 cores aligned, at the same plane, adherent, and easy to cut, with negligible losses during conventional cutting and technical procedures.

Key words: Tissue microarray, TMA construction, Hypodermic needle, Quality control, Immunohistochemistry, Research

1. Introduction

Pathologists have been trying to have a good system to construct multitissue paraffin blocks for more than two decades. The first ingenious reports were from Battifora – "sausage technique" (1) and Wan – "drinking straw technique" (2). 10 years ago, Kononen reported the tissue microarray technique (3).

The conventional construction of a TMA block uses a commercial TMA builder instrument to punch tissue cores from donor blocks and transfer them into holes in a recipient block,

Ronald Simon (ed.), *Tissue Microarrays: Methods and Protocols*, Methods in Molecular Biology, vol. 664,
DOI 10.1007/978-1-60761-806-5_6, © Springer Science+Business Media, LLC 2010

producing blocks with even 1,000 tissue cores. This technique has some disadvantages:

1. To promote adherence between tissue cores and the recipient block, the TMA block has to be lightly heated to melt both; during this phase, cores may misalign;
2. Paraffin fusion may not be complete, resulting in core losses even when using adhesive tape during microtomy;
3. Cores may be inserted too deep in the recipient block, resulting in absence of the most profound cores in the first cuts and;
4. Costs–commercial TMA builder instrument and its disposable needles are not affordable to many pathology laboratories worldwide.

Our alternative technique was developed to produce TMAs without the disadvantages listed above. It is based on the construction of TMA blocks without using a recipient paraffin block for the tissue cores and without using a commercial TMA builder instrument (*4*). Tissue cores are punctured from donor blocks by custom-built hypodermic needles and attached by double-sided adhesive tape on a computer-generated paper grid used to align the cores on the block mold, which is filled with liquid paraffin, creating a block with well-aligned and adherent tissue cores, all at the same cutting plane (Fig. 1). It is an alternative method for the construction of high-density tissue microarray blocks that can be performed by any anatomic pathology laboratory, at low cost and requiring minimum skill and time. It should be emphasized that TMA technique was conceived for use as a research and quality control tool, and not for diagnosis.

2. Materials

2.1. Needles

Conventional hypodermic Becton-Dickinson PrecisionGlide® needles, 16 and 18 gauge (Table 1). Any other needle can be used, like a bone marrow biopsy needle or general core biopsy needles; some of them already have the lateral window (specimen notch).

2.2. Paper Grid

The orientation grid can be done in any drawing software or even in text editors, just using draw, zoom, and alignment tools to create the color background, the circles, and the space between them. The orientation grid displayed here was produced using the software Corel Draw® the following way: 1 mm white circles drawn and aligned leaving 0.3 to 1 mm space between them, on a colorful rectangle background (dark colors are better). The number

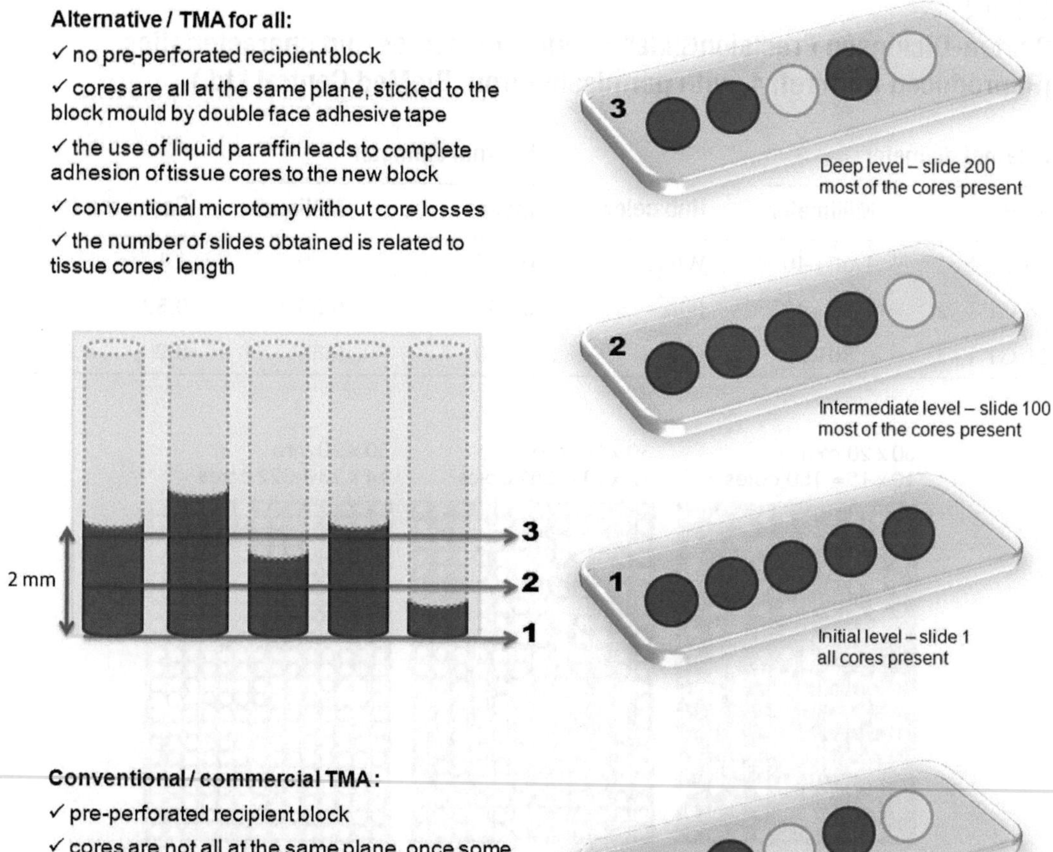

Fig. 1. Schematic comparison between conventional and alternative TMA construction techniques. Note: *black circles* = tissue from core present; *white circles* = no tissue present, *blank space*; *dotted circle* = core lost during microtomy.

Table 1
Becton-Dickinson PrecisionGlide® needles measures and characteristics. (Reproduced from ref. 4, with permission from BioMed Central Ltd.)

External diameter			Internal diameter		
Gauge	Millimeter	Hub color	Inches	Millimeter	Core area (mm^2)
16 Gl 1/2	1.60×40	White	0.047	1.19	1.1
18 Gl 1/2	1.20×40	Pink	0.033	0.83	0.53
21 Gl 1/4	0.80×30	Green	0.020	0.5	0.2

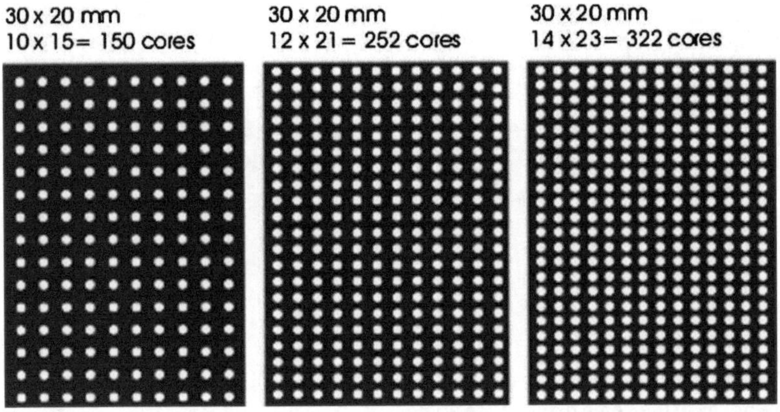

Fig. 2. Paper grids with the same area, but different number of 1 mm circles due to different spaces between circles.

of circles vary considerably by just altering the space between them (Fig. 2). The grid can be printed on plain paper.

2.3. Mold and Double-Sided Adhesive Tape

Paper grids are attached to stainless steel block molds (EasyPath®) by means of a double-sided adhesive tape, 12 mm wide (Scotch 3 M®). The whole upper surface is available for attaching the tissue cores oriented by the grid. The adhesive tape should be wider than the grid, so that its bottom surface free border can attach the paper grid into the mold. Mold size can vary according to the number of cylinders desired (see Notes 1 and 2).

3. Methods

3.1. Needles

Crafting the needles is easy and quick, requiring little skill to produce the new tip and the specimen notch. Everyone is able to produce a good quality needle in less than 15 min after a short period of learning and training (a few hours, consuming 8–10 needles).

Needles, 16 and 18 gauge (Table 1), are prepared for puncturing the donor blocks, as follows:

1. The original bevel is cut off and the tip straightened using a rotary tool system, like Dremel Multi-Pro® Model 395, at low speed (<10,000 rpm), with a cutting wheel attached (Fig. 3a1, b1) (see Note 3).
2. Using the same tool, a lateral opening (specimen notch) is made, 1 mm away from the new tip, with 8–10 mm length and slightly wider than half the needles' width, so that the punched core can be safely removed through it (Fig. 3 c1, a2).
3. The internal surface of the tip is sharpened and imperfections removed from the specimen notch using either a shaft cone stone or a metal sharpener (Fig. 3b2, b3).
4. The external surface of the needle is sharpened with the sanding surface of the cutting wheel, creating a right angle tapered tip – the new cutting edge (Fig. 3a3, b3, C3).

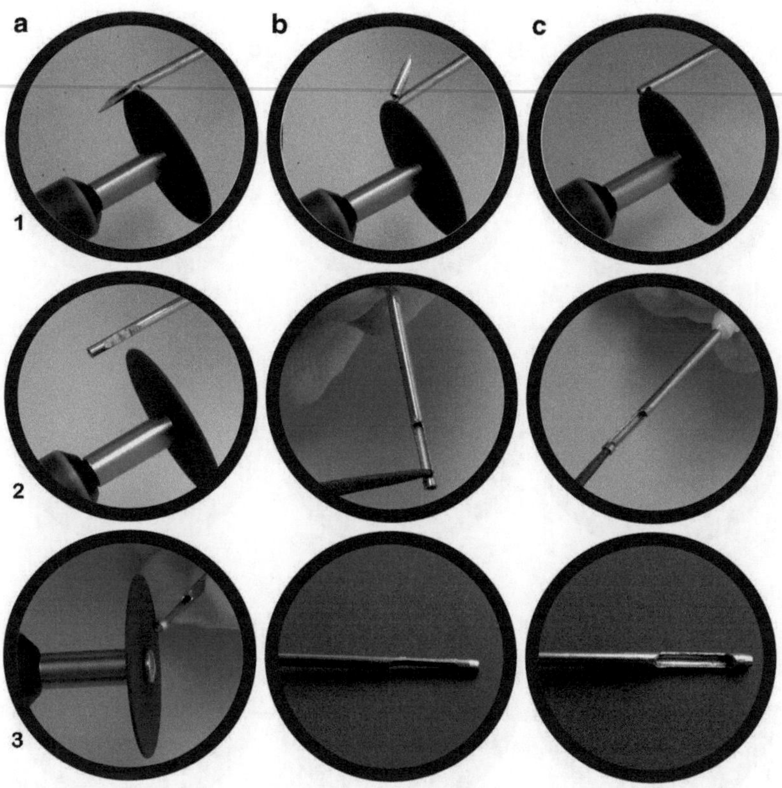

Fig. 3. Needle customization – cutting the original tip (**a1**–**b1**), creating the specimen notch (**c1** and **a2**), sharpening and removing imperfections (**b2**–**a3**) and custom-built needle with specimen notch and new cutting-edge tip (**b3** and **c3**). (Adapted from ref. 4, with permission from BioMed Central Ltd.).

3.2. Puncturing

1. Hematoxylin-Eosin (H&E) slides are used to select the areas to be punctured, which are marked in the slides with permanent marker pen.
2. The paraffin block is overlaid by the slide, the selected areas are recognized, and cores are punctured from the donor blocks.
3. Puncturing can be done manually (Fig. 4a1, b1) or in a semi-automated way, using a hand-press grommet inserting machine with a customized needle attached in the place of the dyes (Fig. 4c1, c2) (see Note 4). It is important that the needle enters the donor block in a perpendicular way, so that the extracted core will have a perfectly cylindrical shape and flat end (see Note 5). This is easier to ensure using the hand-press grommet inserting machine to make the puncture. The specimen notch makes it easier to remove the core with a stylet and tweezers (see Note 6), without harming the tissue (Fig. 4a3, b3). Once hypodermic needles have thin walls and the new cutting edge is sharpened, there is no harm to the donor block, other than the hole created; both 16 and 18 gauge needles yielded the same satisfactory results (Fig. 4c3).

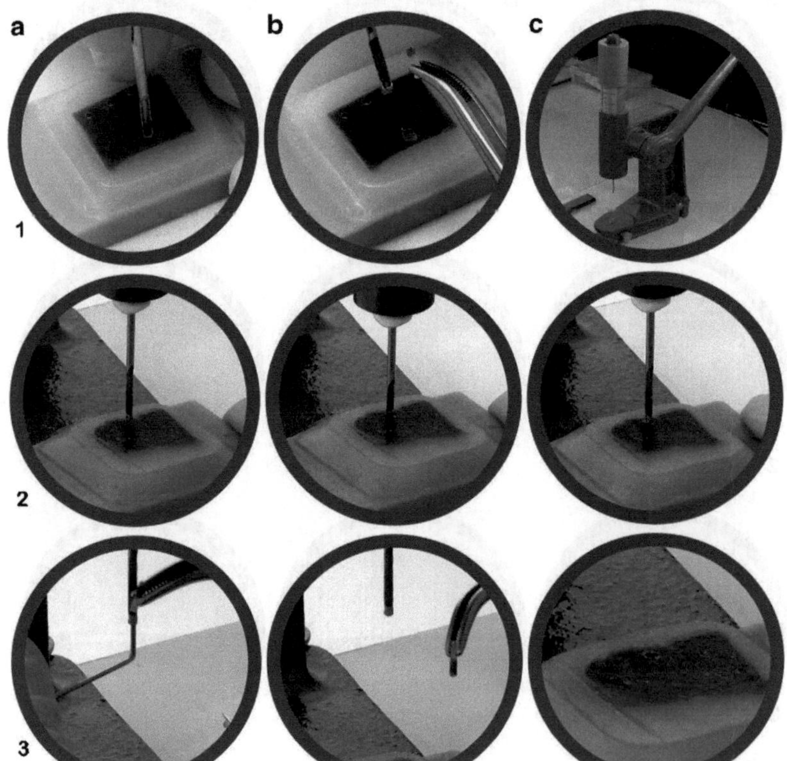

Fig. 4. Paraffin block puncturing, manual (**a1** and **b1**) and with hand-press grommet inserting machine (**c1**–**c3**). (Adapted from ref. 4, with permission from BioMed Central Ltd.).

3.3. TMA Blocks

The construction of TMA blocks using this technique is quick (30 min to make one 25-cores block) and blocks can be handled according to routine histopathological and immunohistochemistry procedures. Sections can be cut on standard rotatory microtome, 3–5 μm thick; each block provides 100–200 slides, depending on the tissue cores' length. There is no need to use adhesive-coated tape-sectioning. There is little core loss (<1%) due to section falling off from slides during technical procedures; most losses are from very thin specimens that end before other thicker cores (see Note 7). The design of each block should be detailed in a TMA map or spreadsheet, indicating the position and identification of each core.

1. Following puncturing, the extracted cores are attached to the double-sided adhesive tape, overlying the selected white circles on the paper grid, in an orderly fashion (Fig. 5a1–c1).
2. Once all cores are attached to the mold, perfectly vertical (Fig. 5a2), melted paraffin is gently poured into the mold (Fig. 5b2), producing a conventional paraffin block with few to hundreds of tissue samples (Fig. 5c2, c3) (see Note 8).

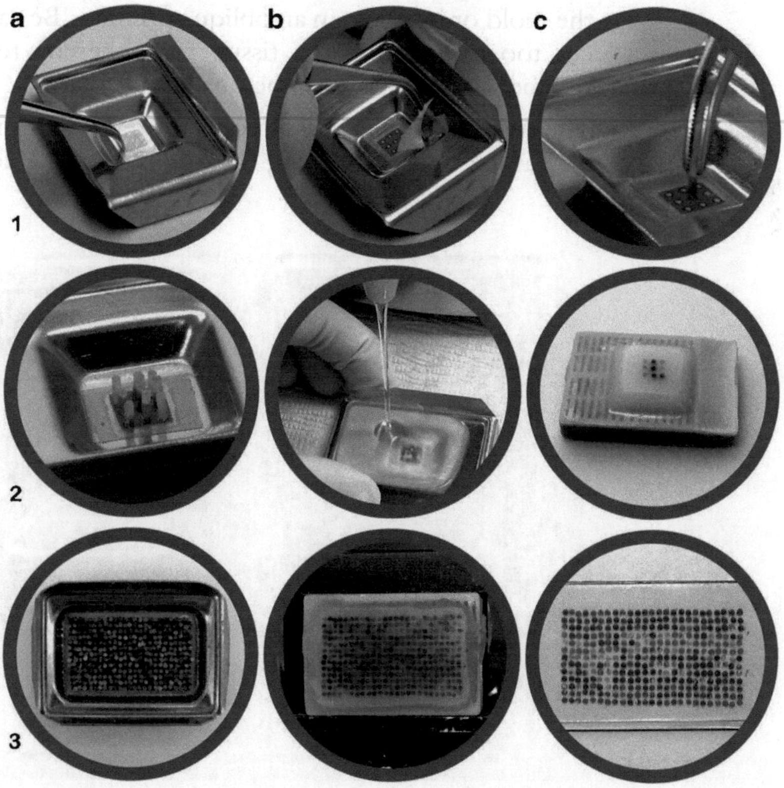

Fig. 5. TMA block production and examples: grid positioning with double-sided adhesive tape (**a1** and **b1**); tissue core being positioned on the grid, over the double-sided adhesive tape (C1); 3 × 3 grid complete with 9 tissue cores in the mold (**a2**); pouring liquid paraffin into the mold (**b2**); TMA block with 9 tissue cores; TMA mold, block, and H&E slide with 325 tissue cores, 25 × 13 grid (**a3–c3**). (Adapted from ref. 4, with permission from BioMed Central Ltd.).

4. Notes

1. It is advisable that the free border of the adhesive tape should be no wider than 2 mm; several variations were tested by us and we noticed that when it was wider, the block could become so glued to the adhesive tape that it could break when trying to get it off from the mold (Fig. 6).
2. It is advisable to use good quality double-sided adhesive tape to guarantee that cores stay upstanding; prefer acetate ones.
3. It is advisable to use protection eyeglasses and gloves during needle customization, in order to avoid harm from metal dust.
4. The total initial cost is near $100.00: Dremel rotatory tool = $40.00; hand-press grommet inserting machine $60.00; hypodermic needles = $0.08; double-sided adhesive tape = $3.00.
5. If the puncture is not perpendicular, the resultant tissue core will have an angled end. It is advisable to cut this angled point with a scalpel and make it plain before placing the core into the mold or it will be in an oblique position. Be careful not to cut too much from the tissue core, thereby reducing the number of slides that can be obtained.
6. The stylet to remove tissue cores from the custom-built needle can be done by cutting the original tip of a conventional

Fig. 6. Paper grids attached to stainless steel block molds by double-sided adhesive tape. *Left*: note long tape strip – this can cause too much adherence between the final TMA paraffin block and the mold, leading to block fissures or fractures. *Right*: adequate tape strip size, 2 mm wider than the paper grid in all sides.

hypodermic needle, 21 gauge, in a perpendicular way and making a right-angle fold 0.5 cm distant from the new tip (Fig. 4a3).

7. It is advisable to avoid donor blocks from which many slides have been cut or that bear too thin tissue specimens (less than 2 mm thick) to prevent irregular consumption of cores, resulting in a reduced number of TMA slides with all cores present.
8. Using grids with very small spaces between circles can lead to difficulty in liquid paraffin flow, producing small bubbles on the block surface. To avoid this, just let liquid paraffin cool slowly; cores will not misalign. If even while doing this a bubble appears, just drop liquid paraffin on it before microtomy.

Acknowledgments

The authors thank Maurício Vincenzi for the photographs. Financial support was provided by Fonte Medicina Diagnóstica Ltda.

References

1. Battifora H. (1986) The multitumor (sausage) tissue block: novel method for immunohistochemical testing. *Lab Invest.* **55**, 244–8.
2. Wan WH, Fortuna MB, Furmanki P. (1987) A rapid and efficient method for testing immunohistochemical reactivity of monoclonal antibodies against multiple tissue samples simultaneously. *J Immunol Methods.* **103**, 121–9.
3. Kononen J, Bubendorf L, Kallioniemi A, Bärlund M, Schraml P, Leighton S, Torhorst J, Mihatsch MJ, Sauter G, Kallioniemi OP. (1998) Tissue microarrays for high-throughput molecular profiling of tumor specimens. *Nature Med.* **4**, 844–7.
4. Pires AR, Andreiuolo Fda M, de Souza SR. (2006) TMA for all: a new method for the construction of tissue microarrays without recipient paraffin block using custom-built needles. Diagn Pathol. **1**:14 doi:10.1186/1746-1596-1-14. The complete electronic version of this article can be found online at: http://www.diagnosticpathology.org/content/1/1/14.

Chapter 7

Resin Technologies: Construction and Staining of Resin TMA's

William J. Howat and Susan J. Wilson

Abstract

The traditional formaldehyde-fixed paraffin-embedded tissue, and therefore the tissue microarrays created from it, provide good morphology but with a compromised antigenicity when compared to frozen tissue. In contrast, while solving the issue of antigenicity, frozen tissue suffers from a lack of morphology. We have demonstrated that tissue microarrays constructed in glycol methacrylate resin, when combined with a cold acetone fixation step, have been able to combine the superior morphology of resin-embedded sections with the superior antigenicity of frozen tissue for prospectively collected material.

Key words: Tissue microarray, Glycol methacrylate, Resin TMA, Immunohistochemistry, Acetone

1. Introduction

Resin has been used as an embedding medium to enable thin sectioning of tissue blocks for electron microscopy since the late 1940s; however, its use in histopathology has been limited to specialised applications, such as bone marrow histochemistry (1) and small biopsy immunohistochemistry (2). Resin has several advantages as an embedding medium over conventional paraffin blocks. Firstly, the density of resin blocks enables a thinner tissue section to be cut, down to 0.5 μm for glycol methacrylate, and thus the optical clarity of the tissue section is only limited by the resolving power of the microscope. Secondly, resin embedding suffers from less tissue shrinkage than paraffin, and thirdly, there is less variability in tissue thickness when cutting resin-embedded material.

Resins can be split into two principal groups: hydrophilic and hydrophobic. The hydrophobic resins, such as Spurr resin (3),

Ronald Simon (ed.), *Tissue Microarrays: Methods and Protocols*, Methods in Molecular Biology, vol. 664, DOI 10.1007/978-1-60761-806-5_7, © Springer Science+Business Media, LLC 2010

remain the predominant resin in use for electron microscopical applications due to their stability in the electron beam; however, their use is limited in immunohistochemistry since the hydrophobicity and highly cross-linked structure of the substrate interferes with the penetration of water-soluble molecules. Therefore, hydrophilic resins are more in use for (immuno) histochemical applications (4, 5). It has generally been shown that embedding in resin does not change the fundamental properties of the tissue, in that with some small modifications, the same applications and stains can be performed following formaldehyde-fixed resin embedding as well as formaldehyde-fixed paraffin embedding (6). However, the combination of acetone fixation and resin embedding for whole tissues, pooled the improved antigenicity of acetone fixation with the improved resolution of resin-embedded thin sections (2, 5) and opened up the possibility of visualising previously unable antigens with superior resin morphology.

The creation of a resin tissue microarray (7) was a natural extension to this technique, but problematic in its construction due to the difficulty in coring tissue already embedded in resin. The application of agarose as a temporary holding structure for the TMA prior to resin embedding, as used in electron microscopy for cell pellets (8), solved this issue and thus resin TMAs could be constructed and their preparation improved.

2. Materials

Acetone is a highly flammable reagent, and its use and therefore TMA construction, should be confined to a fume hood. Glycol methacrylate is a contact sensitising reagent and should be handled using nitrile gloves only. It is sensible to have separate equipment, for example pipettes, for the infiltration and embedding procedure to prevent cross-contamination of the normal lab equipment.

2.1. Fixation

1. "Dry" acetone. Add 25 g of molecular sieve 4A (self indicating) (Fisher Scientific, Loughborough, UK) to 500 ml of acetone. Make up at least 24 h in advance and replace the molecular sieve when granules change colour.
2. Fixation solution. 35 mg of Phenyl Methyl Sulphonyl Fluoride (PMSF, Sigma, Poole, UK) and 370 mg of Iodoacetamide (Sigma) added to 100 ml "dry" acetone. This can be made in advance and stored aliquoted in glass vials at –20°C for up to 1 month.
3. Glass vials with plastic caps (Fisher).
4. Slide labels. Always label in pencil.

2.2. TMA Creation

1. Beecher MTA-1 TMA arrayer (Beecher, Winsconsin, USA) (see Note 1).
2. One 0.6 mm donor coring needle.
3. "Dry" acetone (see step 1 of Subheading 2.1).
4. Fine ridged forceps for gripping tissue.
5. 2% Agarose. Mix 1 g of standard molecular biology grade agarose (Fisher) with 50 ml of RO water. Heat in a microwave with periodic swirling to dissolve the agarose.
6. Tissue Tek metal deep well moulds (Electron Microscopy Sciences, Pennsylvania, USA).
7. Shallow metal container with depth of 1 cm.

2.3. Infiltration and Embedding

1. Infiltration solution. Sufficient JB-4 embedding solution A (Polysciences Inc, Eppelheim, Germany) is precooled for the experiment overnight. 9.5 ml of cooled JB-4 embedding solution A is added to 0.5 ml of methyl benzoate (Fisher Scientific, UK), mixed thoroughly, and stored in a plastic dropper bottle (Fisher) at 4°C when not in use.
2. Embedding solution. 70 mg of Benzoyl peroxide (Polysciences Inc) is added to 10 ml of cooled JB-4 embedding solution A in a plastic dropper bottle. This is shaken gently to dissolve the benzoyl peroxide (*see* Note 2). Once dissolved, 250 µl of GMA solution B (Polysciences Inc) is added, mixed, and the solution used immediately (*see* Note 3).
3. Taab flat-bottomed embedding capsules (Taab Laboratories, Reading, UK).
4. Lined paper, pencil.
5. 8 mm capsule rack (Taab Laboratories).

2.4. Sectioning

1. Semiautomated microtome capable of cutting to 1 µm with glass knife holder, resin block holder attachment, and magnifying lens.
2. 10 mm glass knives prepared using Leica KMR2 glass knife maker (Leica, Milton Keynes, UK).
3. 20 cm × 10 cm Glass trough (RA Lamb, Eastbourne, UK).
4. Fine forceps (Fisher Scientific).
5. Butler block trimmer (Electron Microscopy Sciences).

2.5. Immunohisto chemistry

1. Ventana Discovery automated stainer (Ventana, Tucson, USA).
2. Ventana antibody diluent (Ventana) for dilution of primary and secondary antibodies.
3. DABMap kit.

3. Methods

The methods described refer to the use of the JB-4 embedding kit, which uses the 2-hydroxyethyl methacrylate (also known as glycol methacrylate or GMA) resin. Other hydrophilic resin mixtures, such as methyl methacrylate (9), have also been shown to be suitable for immunohistochemistry.

The cold acetone resin embedding method involves the infiltration of the GMA resin monomer in the first instance alongside methyl benzoate which is used to protect proteins from GMA masking (10). Following infiltration, the tissue is embedded in a mixture of resin, benzoyl peroxide, and N, N dimethylaniline with the benzoyl peroxide acting as a catalyst and the dimethylaniline acting as an accelerator of the system. Mixture of these components drives a free-radical mediated polymerisation of the GMA monomer. This is an exothermic reaction, the temperature is lowered in part by the dimethylaniline, but it is essential that all steps up to and including embedding are conducted at 4°C to minimise cross-linking of the resin and loss of tissue antigenicity. This method is a modification of the JB-4 kit protocol with the addition of less catalyst to lessen resin polymerisation and enhance antigen preservation.

The limitations of the technique should be noted, principally its application to prospectively collected material only, fixed in the correct fixative and difficulty in orientation of specimens that have an epithelial edge, for example, skin.

3.1. Fixation

1. Before starting necropsy, it is helpful to prelabel the glass vials in pencil using standard microscope slide labels, since labelling after will be more difficult and can warm up the acetone and tissue. The cooled acetone fixative must be kept on dry ice during the procedure.
2. Excise each tissue and place into a glass vial containing at least 5 ml of the cooled acetone fixative. After tissue excision, place each individual tissue into a separate glass vial and leave overnight at −20°C (see Note 4).

3.2. TMA Creation

1. Place 0.6 mm donor coring needle in either position of the arrayer (see Note 5).
2. Remove the first five glass vials of tissue for TMA from −20°C storage and leave at room temperature in fume hood for 15 min to warm up. Always loosen the cap of the vials to vent the acetone as it warms, since in some cases the cap can explode off.
3. During this time, pour molten 2% agarose into Tissue-Tek mould to create recipient block and rapid cool on the cool

plate of an embedding station. Lever the agarose recipient block out using flat, slide forceps.

4. Prepare the recipient agarose block for arraying by trimming down excess agarose from rear of block to leave recipient agarose block 2 mm below the level of the recipient block holder. Use the standard grub screws to gently hold the array in place. Do not over-tighten as the array will lift.
5. Add dry acetone to the recipient block holder (see Note 6). Prepare the donor block by filling the shallow metal tray with acetone and placing on the donor block bridge and transferring the first tissue to the metal tray.
6. After 15 min have elapsed, remove the next set of five glass vials from the −20°C storage and set a timer for their 15-min warm-up (*see* Note 7).
7. To create the array, create the first hole in the agarose block then bring the donor block holder into place and core the tissue while holding with ridge forceps. Since there is some compression of less dense tissues, such as lung, during this procedure, it is useful to take multiple, up to five, cores from a single tissue.
8. Repeat with all remaining tissues.
9. Once all cores have been collected, the recipient agarose array can be removed from the recipient block holder and the excess agarose can be trimmed from the sides and bottom of array. This does not need to be too precise since further trimming is required later in the methodology.

3.3. Infiltration and Embedding

1. Immediately following TMA creation, place the TMA/agarose block into a glass vial containing methyl benzoate and incubate at room temperature for 15 min (see Note 8).
2. Using a plastic Pasteur pipette, change the methyl benzoate for 2 ml of the chilled infiltration solution and incubate the vial at 4°C for 2 h. If available, a specimen rotator will improve the infiltration, but is not essential.
3. Remove the "used" infiltration solution and replace with fresh cooled solution using a fresh pipette and incubate for a further 2 h at 4°C.
4. Repeat for a further change of solution at which point the array is ready for embedding (see Note 9). At this point, the array can be tidied up and the top and bottom of the array trimmed to leave no overhanging cores, since the agarose of the array will be completely transparent.
5. Prepare the capsules for embedding; add a thin ring of lined paper (~0.5 × 2 cm) with the array identification number in pencil around the top of the capsule, with writing outermost.

This will set within the resin block. Place the capsules in the capsule rack and add a drop of freshly prepared embedding solution to the capsule.

6. Remove array from infiltration solution and place face down into embedding capsule.
7. Add 1 ml of the freshly prepared embedding solution to the capsule until a meniscus of liquid appears above the level of the capsule and close cap quickly while attempting to minimise the creation of a bubble in the top of the cap (see Note 10). Leave capsule at 4°C for 2 days to complete the polymerisation process.
8. Once polymerised, the cap can be cut off using a razor blade or scalpel and any excess "unused" resin can be removed with a paper tissue.
9. The block must be carefully removed from the capsule using a razor blade and should then be finally dried in a paper tissue before long-term storage at −20°C freezer (see Note 11).

3.4. Sectioning

1. The excess polymerised resin should be carefully trimmed from the back of the block using a razor blade to give a level surface. The edges of the front face of the block must then be trimmed to generate a trapezoid shaped face to aid in rough sectioning on the microtome. A Butler block trimmer is especially useful for this purpose.
2. Place the array in the microtome chuck and rough trim the surface of the block with a fresh glass knife until the full face of the array is being cut.
3. Remove the array, reinsert into the Butler block trimmer and retrim the edges of the array block to a trapezium using a fresh razor blade (see Note 12).
4. Set the microtome to 2-μm sectioning and change to a fresh glass knife. Start the microtome, and using the fine forceps, collect each section as it comes off on the knife (see Note 13).
5. Float sections onto water bath (see Note 14) by tapping forceps on the edge of the bath. Static attraction will otherwise retain the section on the forceps and pick up sections on coated slides. Superfrost Plus, PLL, and APES slides have all been shown to work. Dry at room temperature for 60 min before storing long term at −20°C (see Note 15).
6. Where necessary to stain for morphology, use Mayers Haematoxylin for 2 min followed by 5 min bluing in running tap water, dehydration and mounting in DPX (see Note 16).

3.5. Immunohistochemistry

1. This method assumes the use of the Ventana automated staining system. A manual description of the method is approximated in Note 17.

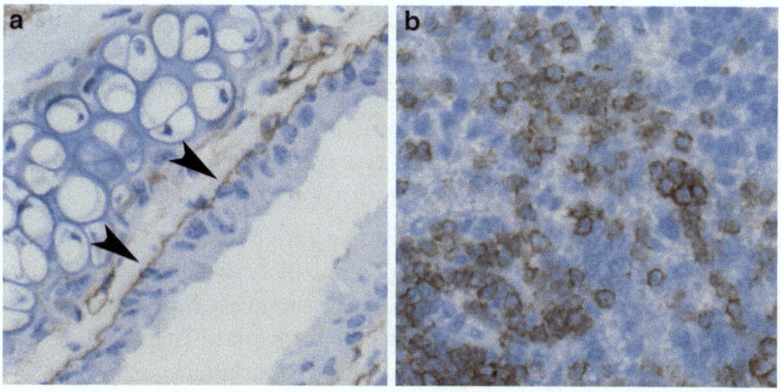

Fig. 1. Cold acetone fixed resin TMA demonstrating: (**a**) Laminin staining of the basement membrane (*arrowheads*) of the murine bronchus, (**b**) CD45 staining of murine spleen.

2. Remove slides from −20°C storage and allow to warm up for a minimum of 30 min before labelling for application to the Ventana system.
3. Any protocol used for formalin fixed-paraffin-embedded tissue will work with the same antibody and the same conditions. Antigen retrieval is necessary when array sections are cut onto water and should be a shortened version of the optimised protocol, for example, CC1 rather than mCC1. Deparaffinisation is not required and a "wet slide load" should be enabled.
4. Antibody incubation times should be adjusted to the maximum allowed, that is, 12 h, to give the antibody time to penetrate the resin. Secondary antibody conditions should also be increased to 2 h. Thereafter, the standard DABMap reagent with a 5-min Haematoxylin will produce crisp positive staining (Fig. 1).

4. Notes

1. The MTA-I arrayer is essential since it has a removable block holder for the donor block. This will contain acetone and will damage any plastic-parts.
2. Do not shake the bottle or mix vigorously, since this induces aeration and will prevent the polymerisation of the resin mixture.
3. This solution will last a maximum of 15 min before discolouration of the reagent will occur, turning yellow/orange. Dispose of the reagent at this stage, since it will not polymerise effectively.
4. Leaving tissue for a longer period in acetone fixative has not been shown to be detrimental to the tissue or the IHC. It is possible to use buffered formaldehyde in place of acetone and treat the resultant array sections in the same way as for

formaldehyde-fixed paraffin-embedded sections. However, formaldehyde-fixed tissue is significantly softer than acetone fixed tissue and more difficult to core.

5. If using lung or less bloody tissues, larger coring needles, such as 1 mm needles, can be used, but will require a longer infiltration procedure.
6. If using an older recipient block holder created by sandwiching a metal plate with a rectangular hole in it with a metal back plate, the acetone will quickly destroy the glue holding the metal back plate in place. Araldite resin can be used to reglue back together.
7. It takes no more than 15 min to core five tissues for the array and this method ensures a continual flow of tissues.
8. This step is a critical requirement for future IHC following acetone fixation.
9. Longer incubations in GMA solutions can be attempted, particularly for larger arrays; however, increasing the time in infiltration solution also increases the GMA-masking effect and should be avoided where possible.
10. The polymerisation process occurs under anaerobic conditions and any bubbles will prevent the curing of the resin.
11. The resin will continue to cure at 4°C and therefore long-term storage at −20°C is essential.
12. A trapezium shape will give a reproducible section that does not skew on the knife, providing sharp edges have been cut into the resin block. It will also enable the block to be placed in the chuck in the same orientation if further sections are cut at a later date.
13. Using a glass knife, a thickness of 2 μm will give reproducible results. For thinner sections, down to 0.5 μm, a Diamond Histo knife (Diatome, Switzerland) can be used. If a glass knife holder is unavailable a tungsten carbide knife can also be used.
14. It is generally recommended in the GMA resin literature that sectioning should be performed on ammonia solution or no IHC result will be obtained (2, 5) (Fig. 2a, b). However, sectioning resin TMAs onto ammonia creates crinkles in individual cores. When sectioning onto water, the crinkles are removed, but antigen retrieval of the array must be performed (Fig. 2c, d).
15. Sections must be used a maximum of 1 month after preparation. Storage at −80°C is not detrimental.
16. Sections can also be stained using 1% toluidine blue plus 1% borax in reverse osmosis (RO) water for approximately 2 min. Wash off excess stain, blot dry and mount in DPX or Pertex.
17. Manual IHC conditions. Most manual IHC conditions will work with resin material with reoptimisation of the antibody

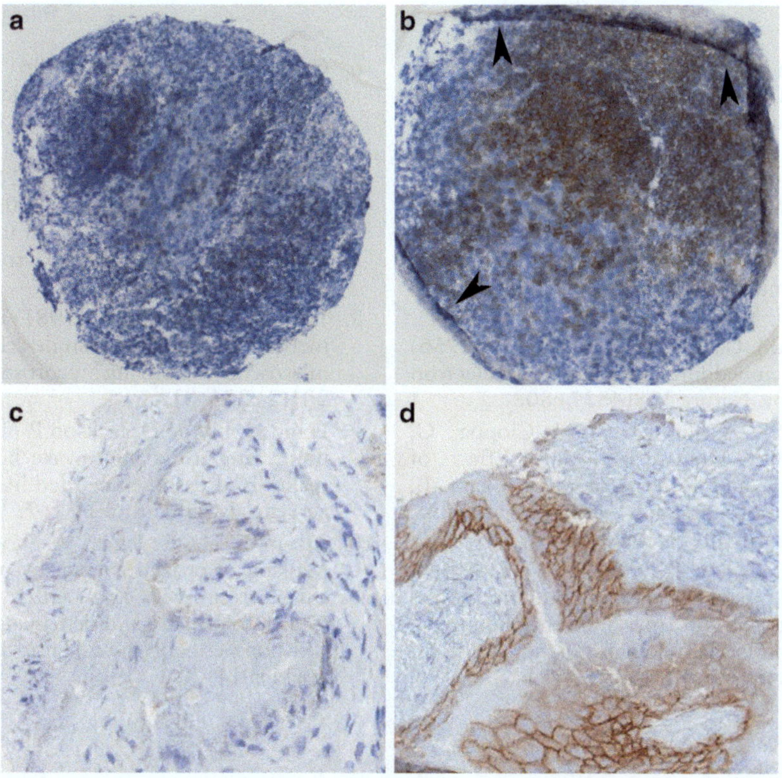

Fig. 2. Resin TMA stained for CD45 on murine spleen (**a**, **b**) or CD138 on murine bladder (**c**, **d**). (**a**) TMA was sectioned onto a water bath and no antigen retrieval was performed. (**b**) TMA was sectioned onto 1% ammonia solution and no antigen retrieval performed. *Note* that there is an increase in staining following floating on ammonia but an imperfect array core is produced with crinkling at the edges (*arrowheads*) (**c**) TMA was sectioned onto a water bath and no antigen retrieval performed. (**d**) TMA was sectioned onto a water bath and heat-mediated antigen retrieval with Tris/EDTA/Borate pH 8 performed. *Note* the increase in staining following antigen retrieval despite the sectioning onto water.

concentrations and antibody incubation times; antibody concentrations may have to be increased and room temperature incubation overnight in a sealed humidified chamber is recommended. To mimic the Ventana antigen retrieval conditions: Protease digestion – alkaline protease, 0.02 U/ml for 10 min at 37°C; Heat mediated antigen retrieval – Tris EDTA Borate, pH 8 or Sodium Citrate pH 6, boiling at 96°C for 8 min and at 100°C for 16 min.

Acknowledgements

The authors would like to acknowledge the technical staff of the immunohistochemistry group of the Atlas of Protein Expression and Histochemistry Research Unit for their contributions to the technical advancements.

References

1. Beckstead JH, Halverson PS, Ries CA, Bainton DF. (1981) Enzyme histochemistry and immunohistochemistry on biopsy specimens of pathologic human bone marrow. Blood 57(6):1088–98.
2. Britten KM, Howarth PH, Roche WR. (1993) Immunohistochemistry on resin sections: a comparison of resin embedding techniques for small mucosal biopsies. Biotech Histochem 68(5):271–80.
3. Glauert AM, Glauert RH, Rogers GE. (1956) A new embedding medium for electron microscopy. Nature 178(4537):803.
4. Wilson SJ, Wallin A, Della-Cioppa G, Sandstrom T, Holgate ST. (2001) Effects of budesonide and formoterol on NF-kappaB, adhesion molecules, and cytokines in asthma. Am J Respir Crit Care Med 164(6):1047–52.
5. Casey TT, Cousar JB, Collins RD. (1988) A simplified plastic embedding and immunohistologic technique for immunophenotypic analysis of human hematopoietic and lymphoid tissues. Am J Pathol 131(2):183–9.
6. Gerrits P, Horobin R. (1996) Glycol methacrylate embedding for light microscopy: Basic principles and trouble-shooting. J Histotechnol 19(4):297–311.
7. Howat WJ, Warford A, Mitchell JN, Clarke KF, Conquer JS, McCafferty J. (2005) Resin tissue microarrays: a universal format for immunohistochemistry. J Histochem Cytochem 53(10): 1189–97.
8. Yuan LC, Gulyas BJ. (1981) An improved method for processing single cells for electron microscopy utilizing agarose. Anat Rec 201(2):273–81.
9. Hand N, Blythe D, Jackson P. (1996) Antigen unmasking using microwave heating on formalin fixed tissue embedded in methyl methacrylate. J Cell Pathol 1:31–7.
10. Takamiya H, Bodemer W, Vogt A. (1978) Masking of protein antigen by modification of amino groups with carbobenzoxychloride (benzyl chloroformate) and demasking by treatment with nonspecific protease. J Histochem Cytochem 26(11):914–20.

Chapter 8

Tissue Microarrays from Frozen Tissues-OCT Technique

Marlena Schoenberg Fejzo and Dennis J. Slamon

Abstract

Tissue microarray technology is a new method used to analyze 100s to 1,000s of tumor samples on a single slide allowing high throughput analysis of genes and proteins on a large cohort. The original methodology involves coring tissues from paraffin-embedded tissue donor blocks and placing them into a single paraffin block. One difficulty with paraffin-embedded tissue relates to antigenic changes in proteins and mRNA degradation induced by the fixation and embedding process. We modified this technology by using frozen tissues embedded in OCT compound as donor samples and arraying the specimens into a recipient OCT block. Tumor tissue is not fixed before embedding, and sections from the array are evaluated without fixation or postfixed according to the appropriate methodology used to analyze a specific gene at the DNA, RNA, and/or protein levels. While paraffin tissue arrays can be problematic for immunohistochemistry and for RNA in situ hybridization analyses, this method allows optimal evaluation by each technique and uniform fixation across the array panel. Frozen tissue microarrays work well for DNA, RNA, and protein analyses, and may have significant advantages over the original technology for the assessment of some genes and proteins by improving both qualitative and quantitative results.

Key words: Frozen, Tissue microarray technology, Immunohistochemistry

1. Introduction

Recently developed high density tissue microarray technology involves arraying up to 1,000 cylindrical tissue cores from individual tumors on a tissue microarray (1). More than 200 serial sections can then be made from an individual microarray block and used for analysis of DNA, RNA, and/or proteins on a single glass slide. The technology is useful in that it allows rapid analysis of a large number of samples so that the statistical relevance of new markers can be determined in a single experiment. In addition, altered expression levels can be correlated to amplification

Ronald Simon (ed.), *Tissue Microarrays: Methods and Protocols*, Methods in Molecular Biology, vol. 664,
DOI 10.1007/978-1-60761-806-5_8,

© Springer Science+Business Media, LLC 2010

or deletion events in specific tumor samples using serial sections, allowing the simultaneous determination of gene copy number and expression analysis of candidate pathogenic genes and suppressor genes. Arrays have been made containing numerous tumor types (2) as well as multiple stages and grades within individual tumor types (3–5). This new technology has already proven useful for rapidly characterizing the prevalence and prognostic significance of differentially expressed genes identified using cDNA array technology (5–7) as well as genes involved in cancer development and progression (4, 5). Tissue microarrays have also been useful in identifying genes that are targets of chromosomal amplification (8, 9) as well as to study the expression patterns of putative tumor suppressor genes (10).

Some technical problems exist with the current methodology, however, relating to the fact that the arrayed samples have been prefixed and embedded in paraffin. The quality of the studies performed on sections from tissue array technology may be limited by the fixation methods used on the original sample. Buffered formalin solutions (and related compounds) are among the most widely used tissue fixatives. These chemicals fix the tissue by acting as progressive cross linkers between proteins and nucleic acids, by introducing modifications in RNA (adding monomethyl groups to its bases), and by producing coordinate bonds for calcium ions; these processes can damage RNA and alter target antigenic structure by blocking or damaging antibody binding sites (11, 12). Formalin fixation-induced alterations can make in situ analysis of DNA, RNA, and proteins suboptimal and variations in the duration of fixation can affect the quality and reproducibility of results (1, 12, 13). Fixation problems for FISH can be overcome by uniformly prefixing tissues in cold ethanol and embedding in paraffin (1) but this approach may not be optimal for array analysis of some proteins or for RNA using in situ hybridization. Paraffin embedding of ethanol-fixed tissue does not prevent RNA degradation (14). In addition, while ethanol fixation of tissue and subsequent paraffin embedding circumvents formalin fixation-related problems introduced by cross-linking, there are still problems relating to the embedding, and/or deparaffinization processes such as temperature-induced antigenic alterations introduced during the embedding process (12, 15, 16). One way to avoid these problems and ensure optimal preservation of antigens and nucleic acids is to use nonfixed (fresh frozen) tissue frozen at –70°C (17, 18). Herein we describe a change in the methodology that circumvents some of the problems with paraffin arrays and demonstrate that frozen tissue is suitable for creating tumor tissue microarrays.

2. Materials

1. Gentle Jane (Instrumedics Inc., Hackensack NJ).
2. Liquid Nitrogen.
3. Dry ice cylinders and sheets.
4. OCT Compound Embedding Medium (Miles, Inc. Diagnostic Division, Elkhart, IN).
5. Cryomold, Standard (Tissue-Tek disposable vinyl specimen molds 25 mm × 20 mm × 5 mm, Miles, Inc. Elkhart, IN).
6. Plastic Paraffin base mold (Simport Histosette II Biopsy Cassette from Fisher Scientific with lid removed).
7. Tissue microarrayer (Beecher Instruments, Silver Spring, MD).
8. Needles (1.0 mm, Beecher Instruments).
9. Basic CryoJane Tape Transfer System and slides (Instrumedics Inc., Hackensack NJ).

3. Methods

The following describes frozen tissue array technology from freezing of samples to using the array sections for IHC, FISH, and RNA in Situ applications. An example of a frozen tissue array is shown in Fig. 1.

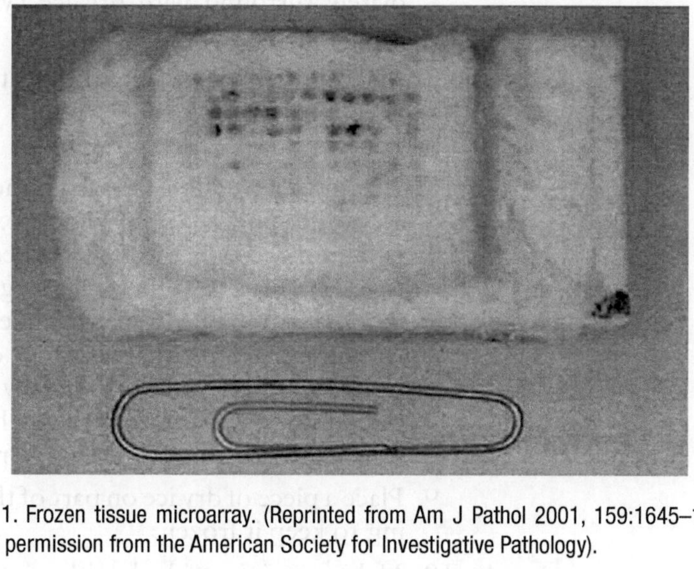

Fig. 1. Frozen tissue microarray. (Reprinted from Am J Pathol 2001, 159:1645–1650 with permission from the American Society for Investigative Pathology).

3.1. Preparation of Samples-Quick-Freezing and Embedding in OCT Compound

1. Procured tissue should be immediately frozen in liquid nitrogen or snap frozen using the Gentle Jane (Instrumedics Inc.) and stored at –70°C (see Note 1).
2. Prior to embedding samples, label molds with freezer pen and pour OCT compound into Cryomolds at room temperature.
3. Remove sample from –70°C freezer and place immediately on dry ice to avoid thawing as much as possible.
4. Pick up sample with prechilled tweezers and quickly push sample down to the bottom until completely submerged in OCT-compound filled Cryomold.
5. Immediately place on dry ice to freeze the embedded sample and avoid thawing as much as possible.
6. Store embedded samples at –70°C.

3.2. Preparation of OCT Recipient Block

1. Fill Cryomold completely and evenly with OCT compound at room temperature.
2. Snap off the base of a plastic paraffin mold, label it with a freezer pen, and make a mark at one corner (for proper orientation when making sections).
3. Place labeled base on top of OCT-filled cryomold and push gently so OCT compound seeps through and can be frozen to the plastic mold.
4. Place on flat surface in –70°C until frozen (approximately 10 min).
5. Once frozen remove plastic Cryomold.
6. Fix OCT Recipient Block in base of tissue arrayer and immediately surround with dry ice cylinders, replacing continuously when evaporated.
7. Make a mark with a freezer pen at 5 mm of your array biopsy needle so that when you punch it you know how deep to punch.
8. Starting a few millimeters from the edge of the block, remove a cylinder punch of OCT (approximately 1 mm in diameter by 5 mm deep) by pushing the needle down into the OCT as far as the mark and then releasing it to pull up the OCT cylinder. The removed OCT cylinder will immediately melt, so wipe the needle with a tissue to clean off the melted OCT, move at least 1 mm horizontally and make the next punch (see Note 2). Continue until you have finished the row. Move down at least 1 mm to make the next row.
9. Place a piece of dry ice on part of the block you are not punching to keep it frozen.
10. Make a recipient block with up to 120 holes in it (12×10) and store at –70°C until samples are ready to array.

3.3. Coring and Arraying Samples

1. Mount the OCT Recipient Block on the tissue arrayer and surround with dry ice cylinders to prevent melting.
2. Place OCT-embedded samples on dry ice. (Remove no more than 10 OCT-embedded samples from −70°C at a time because they will thaw slightly on dry ice).
3. Place OCT-embedded sample on the array bench and biopsy the sample by pushing the needle into the sample using slow pressure and turning the sample slightly from side to side until the needle has reached 5 mm (see Note 3).
4. Remove the biopsy from the sample and immediately hold a piece of dry ice to the needle to keep it from thawing.
5. Immediately return the sample to dry ice and remove the array bench (while keeping a cylinder of dry ice on the biopsy-filled needle).
6. Lower the needle to the appropriate hole on the array and push the biopsy into the hole while keeping a piece of dry ice next to it at all times to avoid thawing.
7. Move 1 mm to the next hole and repeat until the array is filled.
8. Make sure to keep the OCT Recipient block frozen at all times by surrounding continuously with fresh dry ice cylinders and dry ice cylinders on top of the completed rows to avoid thawing (see Notes 4 and 5).

3.4. Sectioning

Sections of the frozen tissue array must be made using the Instrumedics CryoJane Tape-Transfer System.

1. The array is mounted using OCT medium and fixed in the cryostat for sectioning.
2. Make note of the orientation of the array and the orientation of the sections on the slide.
3. Place slides and sectioning tape in the cryostat to cool.
4. Let array warm to the temperature of the cryostat −24 to −32°C.
5. Adjust and section the array until it is smooth and even.
6. Place sectioning tape over the array and use roller to make sure it is evenly sticking to array (see Notes 6 and 7).
7. Cut section.
8. Place section with tape fixed to it on cooled glass slide.
9. Use roller to gently roll section evenly onto slide.
10. Use crosslinker (part of CryoJane Transfer System) to link section to slide.
11. Gently peel off tape leaving section on slide.
12. Place slide with frozen section on dry ice and store slides at −70°C (see Notes 8–10).

4. Notes

1. The quality of the frozen tissue array is highly dependent on how the samples are originally frozen. We did a comparison on freshly removed rat tissues (rat liver tissue shown in Fig. 2) divided into three parts and subsequently frozen (a) in liquid nitrogen, (b) in the Gentle Jane (Instrumedics), and (c) on dry ice. The rat tissues were embedded in OCT, arrayed, sectioned, and HE stained to determine quality. We found tissue arrays made from tissue that was originally frozen more slowly on dry ice to be of poor quality, while tissue originally snap frozen in liquid nitrogen or using the Gentle Jane to be of similarly acceptable quality. Therefore, we find it best to snap freeze the tissue first, then embed the previously frozen sample in OCT compound on dry ice.
2. We prefer to prepare the OCT recipient block by coring holes in it before arraying the samples. This is because it is faster to make holes in the recipient block at a higher temperature when the OCT is a little softer and easier to core. The block can be made simultaneously with the array by coring a hole into the block, inserting the sample, and then coring the next hole, etc., but then care needs to be taken to keep the block completely frozen so the tissue stays frozen.
3. Unlike the paraffin array methodology, the same needle can be used for punching and coring. If a smaller needle is preferred (0.6 mm), be prepared to order extra needles as these needles bend and break easily when forced into frozen samples. The larger needles (1.0 mm or greater) are stronger but pressure should always be applied slowly and carefully to avoid bending or breaking the needle while coring the sample.
4. Even when the sample is surrounded by dry ice cylinders to keep it cold, if the tissue array is not made rapidly, the edges of the array may evaporate off a bit, leaving the array uneven

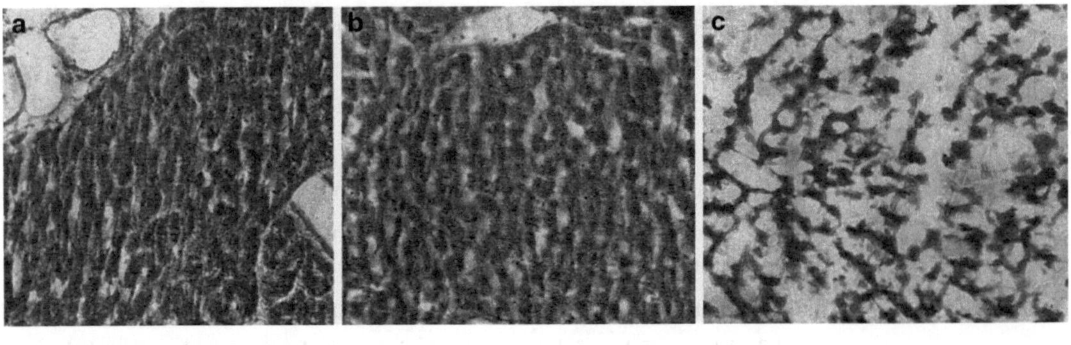

Fig. 2. Freezing methods test.

and difficult to section. Therefore, it is best to center the samples and leave some space around the edges to avoid losing the samples on the edges.

5. The samples cannot be arrayed closer together than 1.0 mm because the OCT can crack when samples are not spaced properly.
6. Unlike the paraffin arrays, you must use the same size needle for preparing the recipient block and coring the sample. Cores that are not fitting tightly into the recipient block can stick to the tape and get pulled out during sectioning.
7. Sectioning can be difficult. The tape and block must be at the correct temperature or the tape will not stick to the block. Also, the tape may not stick to the OCT if it is not smooth, so the block must be cut until it is completely smooth and even, and then beautiful sections can be made. Thus it is important to push the cores all the way into the block when making the array so that you do not lose too much of your samples during the sectioning process.
8. Most importantly, remember to keep the tissue, the array, and the slides frozen at all times to maintain quality until fixation.
9. Remember samples in the frozen tissue array have not been fixed. Before performing IHC, FISH, or RNA in Situ, be sure to fix the array section in the appropriate fixation medium. Probes for FISH and antibodies for IHC are generally more efficient in frozen sections than paraffin, so the dilution of probe or antibody is usually less than recommended by suppliers who have usually optimized their technique for paraffin-embedded samples. For example, for antibodies that recommend 1:100 dilution, a 1:500 dilution may be sufficient.
10. We have also made arrays of frozen cell lines and added cell lines as controls on the tissue array. There are two ways to do this. Cells can be pelleted and snap frozen in a microfuge tube in liquid nitrogen and then removed from the tube with a hammer and quickly embedded in OCT compound similarly to the tissue for subsequent coring. Alternatively, pelleted cells can be pipetted right into the OCT recipient block, but with this method they tend to freeze to the edges leaving a ring of cells with a hole at the center.

References

1. Kononen J, Bubendorf L, Kallioniemi A, Barlund M, Schraml P, Leighton S, Torhorst J, Mihatsch MJ, Sauter G, Kallioniemi OP (1998): Tissue microarrays for high-throughput molecular profiling of tumor specimens. Nat Med, 4:844–847.
2. Schraml P, Kononen J, Bubendorf L, Moch H, Bissig H, Nocito A, Mihatsch MJ, Kallioniemi OP, Sauter G (1999): Tissue microarrays for gene amplification surveys in many different tumor types. Clin Cancer Res, 5:1966–1975.

3. Moch H, Schraml P, Bubendorf L, Mirlacher M, Kononen J, Gasser T, Mihatsch MJ, Kallioniemi OP, Sauter G (1999): High-throughput tissue microarray analysis to evaluate genes uncovered by cDNA microarray screening in renal cell carcinoma. Am J Pathol, 154:981–986.
4. Bubendorf L, Kononen J, Koivisto P, Schraml P, Moch H, Gasser TC, Willi N, Mihatsch MJ, Sauter G, Kallioniemi OP (1999): Survey of gene amplifications during prostate cancer progression by high-throughout fluorescence in situ hybridization on tissue microarrays. Cancer Res, 59:803–806.
5. Bubendorf L, Kolmer M, Kononen J, Koivisto P, Mousses S, Chen Y, Mahlamaki E, Schraml P, Moch H, Willi N, Elkahloun AG, Pretlow TG, Gasser TC, Mihatsch MJ, Sauter G, Kallioniemi OP (1999): Hormone therapy failure in human prostate cancer: analysis by complementary DNA and tissue microarrays. J Natl Cancer Inst, 91:1758–1764.
6. Moch H, Schraml P, Bubendorf L, Mirlacher M, Kononen J, Gasser T, Mihatsch MJ, Kallioniemi OP, Sauter G (1999): Identification of prognostic parameters for renal cell carcinoma by cDNA arrays and cell chips. Verh Dtsch Ges Pathol, 83:225–232.
7. Barlund M, Forozan F, Kononen J, Bubendorf L, Chen Y, Bittner ML, Torhorst J, Haas P, Bucher C, Sauter G, Kallioniemi OP, Kallioniemi A (2000): Detecting activation of ribosomal protein 86 kinase by complementary DNA and tissue microarray analysis. J Natl Cancer Inst, 92:1252–1259.
8. Barlund M, Monni O, Kononen J, Cornelison R, Torhorst J, Sauter G, Kallioniemi OP, Kallioniemi A (2000): Multiple genes at 17q23 undergo amplification and overexpression in breast cancer. Cancer Res, 60:5340–5344.
9. Richter J, Wagner U, Kononen J, Fijan A, Bruderer J, Schmid U, Ackermann D, Maurer R, Alund G, Knonagel H, Rist M, Wilber K, Anabitarte M, Hering F, Hardmeier T, Schonenberger A, Flury R, Jager P, Fehr JL, Schraml P, Moch H, Mihatsch MJ, Gasser T, Kallioniemi OP, Sauter G (2000): High-throughput tissue microarray analysis of cyclin E gene amplification and overexpression in urinary bladder cancer. Am J Pathol, 157:787–794.
10. Bowen C, Bubendorf L, Voeller HJ, Slack R, Willi N, Sauter G, Gasser TC, Koivisto P, Lack EE, Kononen J, Kallioniemi OP, Gelmann EP (2000): Loss of NKX3.1 expression in human prostate cancers correlates with tumor progression. Cancer Res, 60:6111–6115.
11. Masuda N, Ohnishi T, Kawamoto S, Monden M, Okubo K (1999): Analysis of chemical modification of RNA from formalin-fixed samples and optimization of molecular biology applications for such samples. Nucleic Acids Res, 27:4436–4443.
12. Werner M, Chott A, Fabiano A, Battifora H (2000): Effect of formalin tissue fixation and processing on immunohistochemistry. Am J Surg Pathol, 24:1016–1019.
13. Specht K, Richter T, Muller U, Walch A, Werner M, Hofler H (2001): Quantitative gene expression analysis in microdissected archival formalin-fixed and paraffin-embedded tumor tissue. Am J Pathol, 158:419–429.
14. Goldsworthy SM, Stockton PS, Trempus CS, Foley JF, Maronpot RR (1999): Effects of fixation on RNA extraction and amplification from laser capture microdissected tissue. Mol Carcinog, 25:86–91.
15. Battifora H, Kopinski M (1986): The influence of protease digestion and duration of fixation on the immunostaining of keratins: a comparison of formalin and ethanol fixation. J Histochem Cytochem, 34:1095–1100.
16. Penault-Llorca F, Adelaide J, Houvenaeghel G, Hassoun J, Birnbaum D, Jacquemier J (1994): Optimization of immunohistochemical detection of ERBB2 in human breast cancer: impact of fixation. J Pathol, 173:65–75.
17. Slamon DJ, Godolphin W, Jones LA, Holt JA, Wong SG, Keith DE, Levin WJ, Stuart SG, Udove J, Ullrich A, Press MF (1989): Studies of the HER-2/neu proto-oncogene in human breast and ovarian cancer. Science, 244:707–712.
18. Battifora H (1991): Effect of fixatives and fixation times on tissues. Am J Clin Pathol, 96:144–145.

Chapter 9

An Alternative Technology to Prepare Tissue Microarray Using Frozen Tissue Samples

Zhongting Hu, Elbert Chang, and Melissa Hodeib

Abstract

Although most tissue microarray (TMA) slides are currently made from paraffin-embedded tissues, frozen clinical tissues are also gradually being used to prepare TMAs. This is because frozen tissues contain better quality RNAs and proteins for profiling gene expressions. Here, we introduce another TMA method that is applicable to a broader range of frozen tissue samples.

In this method, an agarose-gel-based array recipient block is first made using several simple instruments. Frozen donor tissues are then manually cored and arrayed into the recipient block array at −10°C. After arraying, the array block can be immediately sectioned on a cryostat microtome to make TMA slides for in situ hybridization and immunocytochemistry studies. TMAs made by this method have well-defined array configurations, good tissue/cell morphology, and well-preserved proteins and mRNAs. This low-cost and time-saving method provides an alternative tool for preparing high quality TMAs for gene expression analyses.

Key words: Recipient block, Frozen tissues, Coring, Tissue arraying, mRNAs, Proteins

1. Introduction

Although most TMAs are made from paraffin-embedded tissues (1), some researchers prefer to use frozen clinical tissues to prepare TMAs (2, 3). This is because fresh frozen tissues provide better preserved RNAs and proteins for gene profiling analysis. Recently, an optimal cutting temperature compound (OCT)-embedding method (4) was developed to make TMAs from frozen tissues. In this approach, frozen tissues are embedded in OCT without fixation. The embedded tissues are cored and transferred into an OCT recipient block in dry ice. The OCT-based method has been described by Fejzo et al. in this book series.

Ronald Simon (ed.), *Tissue Microarrays: Methods and Protocols*, Methods in Molecular Biology, vol. 664, DOI 10.1007/978-1-60761-806-5_9, © Springer Science+Business Media, LLC 2010

Here and elsewhere (5), we describe an alternative method to prepare TMAs from frozen tissue samples. This method involves the use of several custom-made instruments to create an agarose gel-based recipient block that contains preformed array cavities. This recipient block is fundamentally different in design and composition from both paraffin-recipient blocks (1, 6) and OCT blocks (4). This recipient block can be used to manually array tissue cores at −10°C inside a cryostat microtome or a cold room. After arraying frozen tissue cores into the recipient block, the tissue block can be sectioned immediately in a −20°C cryostat microtome. The mounted TMA slides can be used immediately for mRNA in situ hybridization and immunocytochemistry studies. TMAs made by this method maintain the integrity of RNAs and proteins, which is particularly important for profiling gene expressions in clinical tissue samples. Our method provides an alternative low-cost and time-saving method to make high quality TMAs from frozen tissues.

2. Materials

2.1. Instrumental Components for Preparing an Array Recipient Block

The agarose-gel-based array recipient block is made with several custom-made tools: a metal stamp, a plastic mold box, and a gel-retaining plate (Fig. 1).

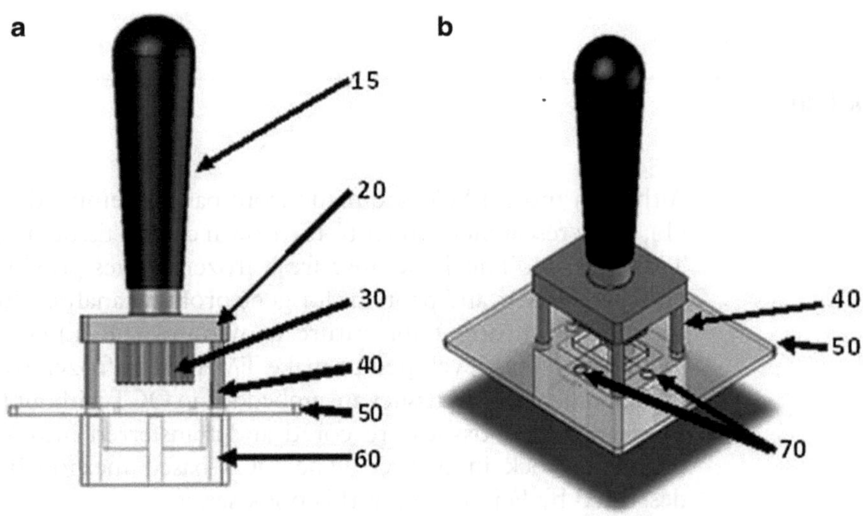

Fig. 1. Illustrations show two views of the instruments used for producing array recipient blocks. Array-forming stamp comprises a handle (15), a support base (20) attached to 96 pins (30), and guiding pins (40). The mold box (60) is vapor-polished to promote see-through clarity. It contains a deep pocket in the geometric *center* of the *top* of the box. The gel-retaining plate (50) is also vapor-polished to promote see-through clarity. It contains a rectangular hole located in the geometric center and is secured to the mold box by four locating pins (70).

1. *Array-forming stamp*: The steel stamp is used to form cylindrical array cavities in the recipient block. The stamp comprises a plastic handle (90 mm long and 26 mm in diameter narrowed at its attachment point), a support base (53 × 43 × 10 mm) with holes to accommodate 96 pins, and four guiding pins (5 mm in diameter and 32 mm long). The support base is fabricated from stainless steel. It is 53 × 43 × 10 mm, and contains a boss 10 mm tall × 19 mm in diameter and a female thread centered on the top portion of the boss for accommodating the handle. The plastic handle contains a threaded stud insert approximately 13 mm in length located at the narrow end of the handle. This allows the handle to be threaded onto female threads of the support base. Each of the 96 steel pins is 1.5 mm in diameter and 22 mm long. They are standard hardened stainless steel dowel pins and press-fit into the corresponding holes of the support base. The 96 pins have a uniform center-to-center distance of 1 mm.
2. *Plastic mold box*: The polycarbonate cubic box (53 × 43 × 26 mm) holds the dissolved array recipient matrix and the 8 × 12 precisely arrayed pins on the steel stamp form the array cavities as the gel matrix solidifies. It is vapor-polished with a wall thickness of 8 mm and a center pocket (37 × 28 × 13 mm). It also contains vertical hollow shafts (5 mm diameter) at each corner to accommodate guiding pins and additional shafts within its four walls to secure the locating pins of the gel-retaining plate.
3. *Gel retaining plate*: The polycarbonate gel-retaining plate (104 × 95 × 3 mm) is used to keep the array recipient block in the plastic mold box. It is vapor-polished and contains a rectangular hole (28 × 19 mm) in its center. It is fused to four locating pins (6.5 mm long and 5 mm in diameter) positioned symmetrically about the main center of the hole and spaced 45 × 35 mm apart. They are designed to fit precisely into the corresponding hollow shafts in the plastic mold box. The gel-retaining plate also contains four holes (5 mm in diameter) located symmetrically about the main center of the hole and spaced 45 × 35 mm apart that are designed to accommodate the guiding pins of the stamp.

2.2. Instrument for Coring Tissue and Transferring Samples

A cylindrical thin-walled coring tube (5.4 cm long and 1.53 mm inner diameter) and plunger (5.5 cm long and 1.5 mm diameter) together referred to as a coring borer, (Fig. 2a, b) are used for coring tissues and transferring cored samples into array cavities in the recipient block. The inner diameter of the coring tube is slightly larger than the array cavity, allowing the cored tissue to fit snugly inside the cavity to facilitate tissue sectioning. The coring tube has

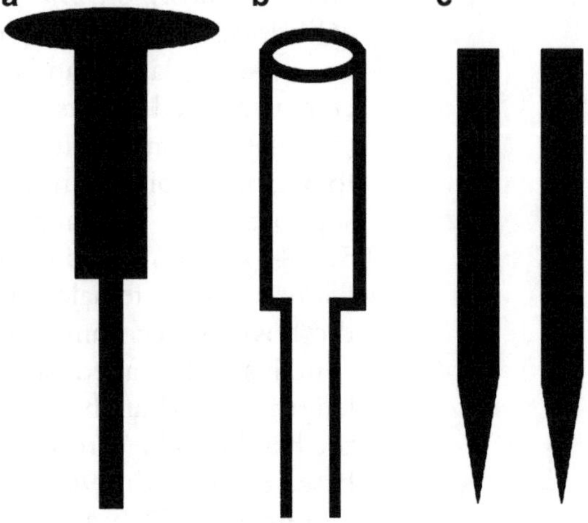

Fig. 2. Sketch shows the coring borer (**a**, **b**) and the design of the tip (**c**) of the coring tube. The coring borer consists of two parts: the plunger (**a**) and the coring tube (**b**). The plunger can be inserted into the tube to push the cored tissue out. Different types of tips have been tested. Only the tube with inside and out chamfer (**c**) yields cored tissue samples with identical cylindrical shape.

a flanged inner tip chamfer (Fig. 2c) design so that the cored tissues have uniform cylindrical shapes.

2.3. Materials for Reparation of Premade Array Recipient Block

1. Agarose gel (agarose-HR) separates DNA fragments from 50 to 1,000 bp. E.E.O, Gel strength 1.5%, >750 gm/cm^2 (Sigma St. Louis, USA).
2. Sucrose (Sigma USA, Cat#: S-7903).
3. Four percent paraformaldehyde solution stored at room temperature (Sigma, St. Louis, USA).
4. 1× TBE, 10.8 g Tris Base, 5.5 g Boric acid and 0.93 g Na4EDTA in 1 L of distilled water.
5. Gel matrix solution (11.5% sucrose, 6.0% agarose, 1 ml of 8% paraformaldehyde, 1× TBE).

2.4. Other Reagents and Materials

1. Optimal cutting temperature compound (OCT) (VWR, Cat#: 25608-930).
2. Washing solution: 0.5% (v/v) Triton X-100 in 1× PBS solution.
3. 100% Ethanol.
4. Dry ice.
5. Fixed or unfixed frozen normal or pathological tissue samples.

2.5. Other Instruments for Making TMA Slides

1. Cryostat microtome.
2. SuperFrost microslides (VWR, Cat#48311-703).

3. Methods

3.1. Preparation of an Agarose-Gel-Based Recipient Block (50 ml)

1. In a 200-ml Erlenmeyer glass flask, prepare 50 ml of gel matrix solution.
2. Stir the solution 1–2 min at room temperature (RT) to dissolve the sucrose.
3. Place the flask in a microwave to melt the agarose and then let the solution stand at RT for a few minutes to eliminate air bubbles from the solution.
4. Pour the gel matrix solution into the plastic mold box, place the gel-retaining plate on the box, and then slowly insert the array-forming stamp into the plastic mold box (see Note 1).
5. Leave the gel at RT for about 15–20 min to let the gel matrix solidify.
6. Remove the array-forming stamp from the gel while holding the gel-retaining plate firmly to prevent the array recipient block from being torn. Multiple cylindrical array cavities will be formed inside the array recipient block (Fig. 3).
7. Carefully remove the array recipient block from the box. Wrap the array recipient block with Parafilm to maintain moisture of the array recipient block. The wrapped block can be stored at RT or 4°C for several months.

3.2. Tissue Preparation for Making TMAs

Both fresh and archival frozen tissues (fixed or unfixed) can be used to make TMAs.

1. Place frozen tissues (kept at –80°C or on dry ice) in a –20°C freezer for 30 min to 1 h. This will increase their temperature so that they can be easily cut into desired sizes.

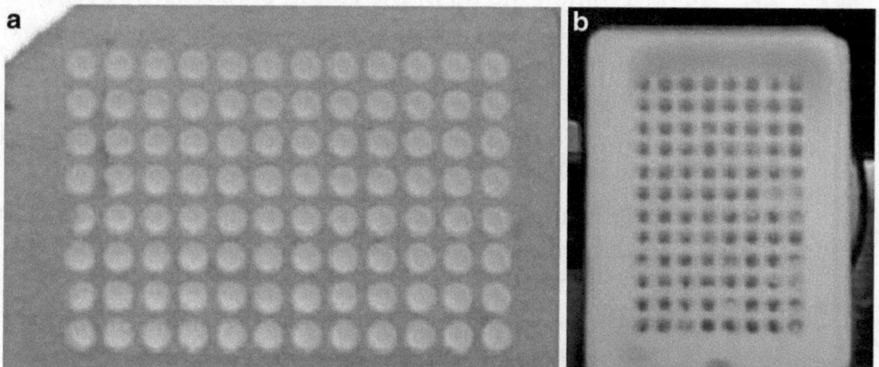

Fig. 3. Examples of a 96-well array recipient block. *Left hand panel* is the array recipient block at room temperature and *right hand panel* is the array recipient block frozen by dry ice.

2. Cut large tissues into 1 × 1 × 1 cm cubes. Small tissue samples can be used directly.
3. On dry ice, put the tissue samples into small weight boats (2 × 2 × 1 cm; one tissue sample per weight boat). Label the tissue type on the back side of the weight boat.
4. Pour OCT compound into the weight boat, submerging the tissue samples completely, and then quickly freeze the OCT by covering with dry ice for 10–15 min. The tissue samples and embedding medium should turn completely white after 15 min in dry ice (see Note 2).
5. Mark the position of embedded tissues by punching a small hole in the O.C.T so that the coring site can be easily identified. Now, the tissue samples are ready to be cored.
6. Quickly freeze array recipient block in dry ice for 15 min (see Note 3).

3.3. Preparing Tissue Microarray Block

1. Move the frozen tissue samples, array recipient block, and coring borer into −10 to −15°C cryostat microtome for about 30 min to equilibrate the temperature to −10 to −15°C. At this temperature, frozen tissue samples can be easily cored (see Note 4).
2. Place the tissue sample on the stage of the cryostate microtome.
3. Mark the orientation of the array recipient block by making a small cut in one corner.
4. Remove plastic plunger from coring tube. Use the coring tube to core tissue samples. Coring is best done using a steady and slow motion with gentle pressure.
5. Make sure the coring tube goes through the tissue samples. Check the tip of coring borer to confirm tissue sample is inside and in good round shape.
6. Pull the coring tube out of the tissue samples.
7. Position the tip of the coring borer exactly over an array cavity, place the plastic plunger back into the coring tube, and then depress the plastic plunger gently and continuously to push the cored tissue sample into the cavity.
8. Record the position of the arrayed tissue samples in the recipient block in a notebook.
9. Repeat steps 4–8 to array all the prepared tissue samples into the array recipient block. It usually takes under 2 h to array 96 samples.
10. After arraying, the TMA block can be cut immediately or stored at −80°C until needed (see Note 5).
11. Lower the temperature of cryostat microtome to −20°C. Trim the recipient block to fit the cutting knife of the cryostat microtome if necessary.

12. Adjust the cutting angle of the knife and position the anti-rolling glass plate to prevent the curling of tissue sections during cutting. A TMA section can be cut into 5–16 μm sections (see Notes 6 and 7).
13. Transfer tissue sections onto precoated glass slides and air-dry them for 10–15 min at RT (see Notes 8 and 9). The TMA slides are then ready for analysis. Stain one slide with H&E to check the overall configuration and tissue morphology (see Note 10). This method consistently yields high quality TMAs with good core arrangement and good tissue and cell morphology (Fig. 4). Extra TMA slides can also be stored at −70°C for future use.

3.4. Pretreat Tissue Microarray Slides Before Experiment

1. Air dry for 10 min at RT or bake TMA slides at 37°C for 10 min to allow the TMA sections to stick to the glass slides.
2. Dip the slides in 4% paraformaldehyde containing 1× PBS for 10–15 min to fix tissue cores if the tissue samples have not been fixed previously (see Note 11).
3. After fixation, wash the TMA slides with 1× PBS for 5 min twice. This wash will usually remove the array matrix from the slides. Array sections are now ready for in situ and immunohistochemistry studies (see Notes 12–14).

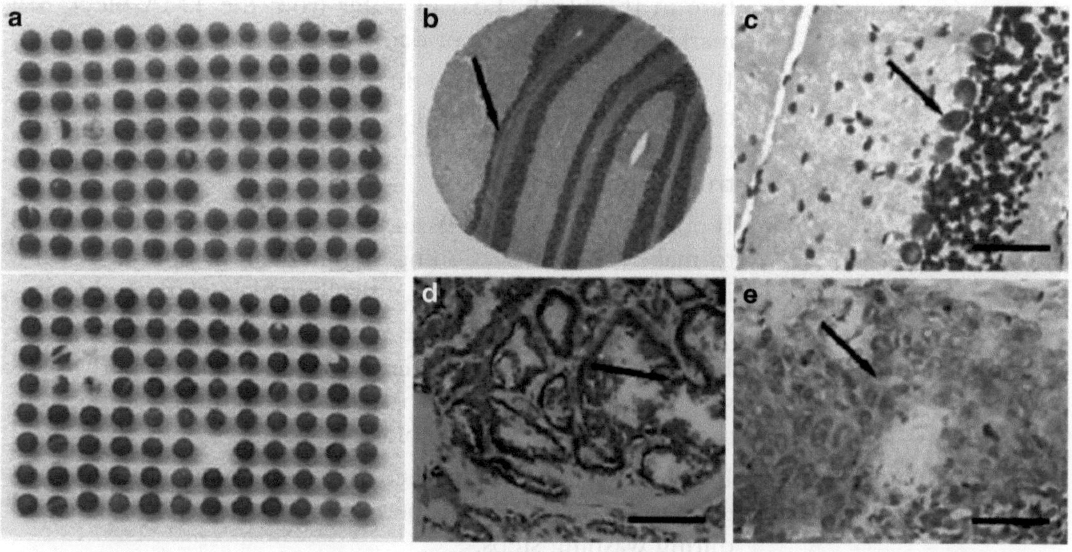

Fig. 4. Histological staining of TMA slides. All the cores are well aligned as revealed by cresyl violet (*upper panel* in **a**) and H&E staining (*lower panel* in **a**). In a cresyl violet-stained mouse cerebellum (**b**), the edge of the core is sharp and the granule cell layer (*arrow* in **b**) is clearly present. *Panel* **c** shows the Purkinje cells (*arrow*) and granule cells in the mouse cerebellum. Neither empty holes nor separation of tissue are present. *Panel* **d** shows the structure of pancreatic cancer tissue. No freeze/thaw-induced tissue debris is observed. In a frozen lung cancer sample, the typical nuclear profiles of cancer cells (*arrow* in **e**) are present, large and irregularly shaped. Bar in $C=50$ μm, in $D=200$ μm, in $E=150$ μm.

4. Notes

1. No air bubbles should be present inside the gel when the stamp is inserted.
2. It is very important to freeze tissues/embedded tissues quickly. Otherwise, tissue physical properties and morphology will be affected.
3. Rapid freezing of the array recipient block is very important. Otherwise, the array recipient block cannot be sectioned smoothly.
4. If the tissue samples are hard to core, it is usually because the temperature of sample is too low. Samples from dry ice or −80°C need to be equilibrated to −10 to −15°C for 30 min in the cryostate microtome before coring. If the temperature of the sample is too low, the sample may also shatter during coring. If the tissue sample is too soft to core, the temperature of the tissue sample may be too high. Monitor the temperature of the cryostat microtome or cold room by placing a thermometer inside the microtome or cold room. The tissue and cell morphology may be compromised if the temperature of the tissue samples is higher than −5°C.
5. Before storing in −80°C freezer, wrap the TMA block with parafilm and then place it in an air-tight box to retain moisture in the block. Loss of water from the TMA block may markedly compromise or damage tissue morphology.
6. If TMA section cores curl during cutting, make sure the anti-rolling glass plate is placed at the optimal position. The plate must be positioned parallel with the cutting knife and protrude slightly beyond it.
7. If possible, cut normal tissue the same size as the array block to make sure the sectioning system is at an optimal setting. Keep the cutting knife and antirolling glass plate cool to prevent sections from adhering to the antirolling glass plate.
8. If tissue cores do not transfer to the glass slide properly, make sure that the slides are coated with coating material. Use slides recommended in Subheading 2.5.
9. If the glass slides are not coated properly or are not completely dry before the experiment, tissue cores will fall off during washing steps.
10. Due to variations in cellularity, connective tissue and fat content, small holes and gaps might be present in some tissues (e.g., muscle and spleen).
11. This fixation will quickly inhibit RNases and other enzymes to protect the integrity of RNAs and proteins in the tissue cores.

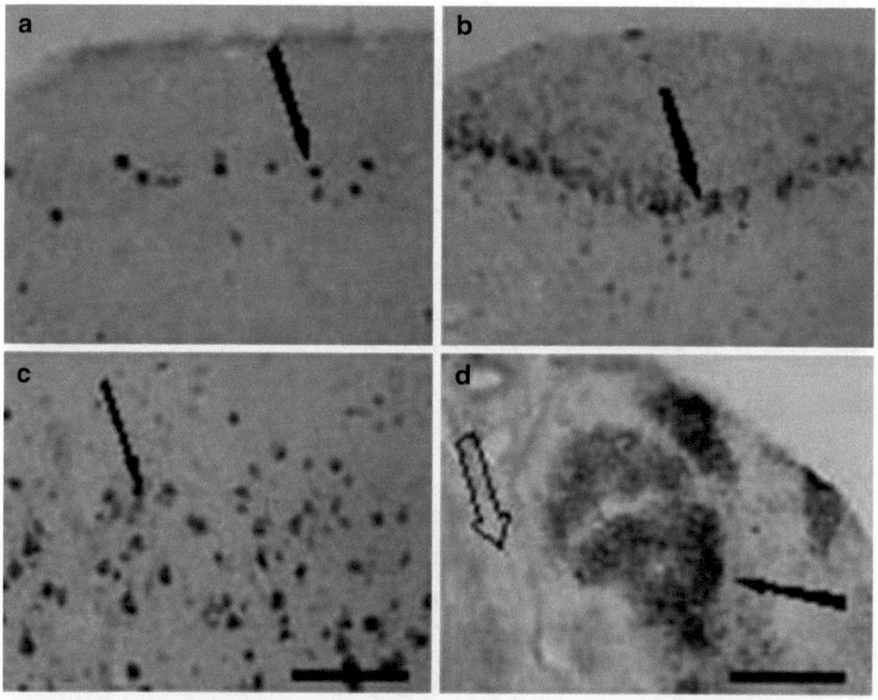

Fig. 5. Illustrations show expressions of VIP and GAPDH mRNA in tissue cores (*solid arrows* indicate hybridized cells): VIP mRNAs in mouse cerebral cortex (**a**) GAPDH mRNAs in the mouse Purkinje cell layer (**b**) and human cerebral cortex (**c**). In human ovarian cancer tissues, the GAPDH mRNA (*solid arrow* in **d**) was more abundant in cancer cells than in adjacent normal cells (*open arrow* in **d**). Scale Bar *C*=50 μm (for **a**–**c**), *D*=500 μm. Hybridization was done according to published protocol (7).

12. Low RNA signals during in situ hybridization can result from long and improper storage of the tissues samples. Cut original donor tissues to check the RNA quality and the abundance of the target mRNA samples in the experimental tissues. This method usually produces TMA slides with high quality tissue mRNA (Fig. 5).
13. Low protein signal during immunocytochemistry could be the result of antigenic change from the improper preservation of tissue samples. Cut original donor tissues to check the protein quality and also check the abundance of target proteins in experimental tissues. TMAs made by this method usually contain high quality proteins for experiment (Fig. 6). Immunocytochemistry was done based on a published protocol (8).
14. If the array matrix does not fall off the glass slide after the PBS washes, it will be washed away during the washing steps in your immunocytochemistry and in situ hybridization experiments.

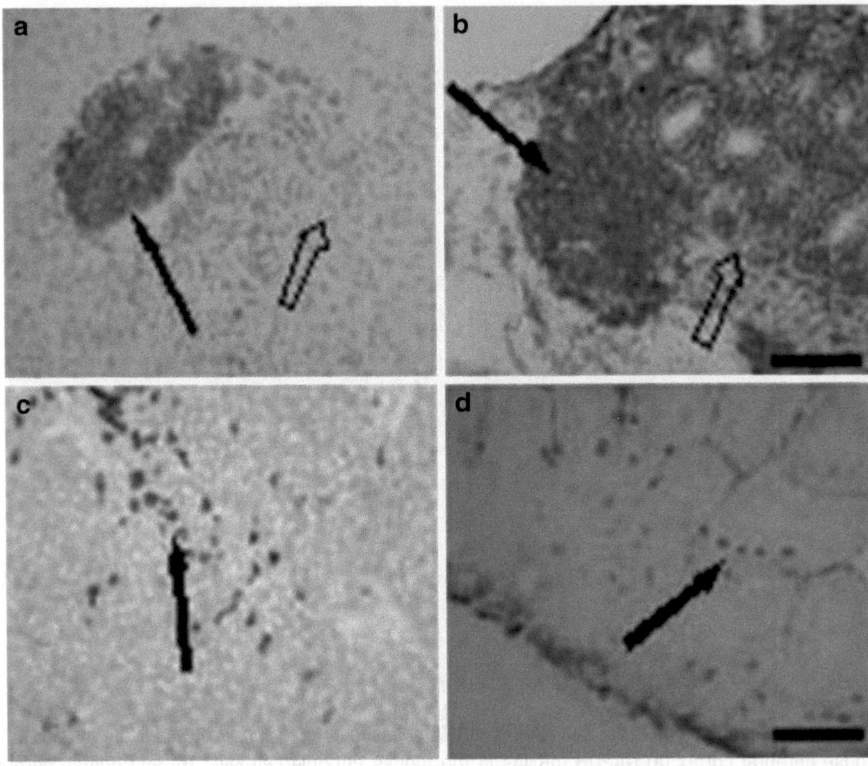

Fig. 6. Counter-stained slides show expressions of some proteins in prepared TMA cores. Immunoreactive material is indicated by dark grey color. Strong GAPDH expression is seen in ovarian cancer tissue (*solid arrow* in **a**). The normal adjacent cells have weak GAPDH (*open arrow* in **a**). *Panel* **b** shows cytokeratin in colon cancer tissue (*solid arrow* in **b**). *Open arrow* in (**b**) indicates the lack of cytokeratin in adjacent cell region. Ras cells (*arrow* in **c**) are clearly detected in lung cancer tissue. In mouse intestine, CHAT cells are also clearly identified (*arrow* in **d**). Scale bar in *B* = 100 μm (for **a**, **b**) and in *D* = 100 μm (for **c**, **d**).

Acknowledgments

We would like to thank Dr. Amy Chew for comments. This research is supported by NIH grants 1R43GM074318-01 and 1R44CA119651-01, to Z. Hu.

References

1. Kononen, J., L. Bubendorf, A. Kallioniemi, M. Barlund, P. Schraml, S. Leighton, J. Torhorst, M.J. Mihatsch, et al. (1998). Tissue microarrays for high-throughput molecular profiling of tumor. Nat Med 4:844–847.
2. Scicchitano, M.S., D.A. Dalmas, M.A. Bertiaux, S.M. Anderson, L.R. Turner, R.A. Thomas, R. Mirable, R.W. Boyce (2006). Preliminary comparison of quantity, quality, and microarray performance of RNA extracted from formalin-fixed, paraffin-embedded, and unfixed frozen tissue samples. J Histochem Cytochem 54:1229–1237.
3. Hoos, A., C. Cordon-Cardo (2001). Tissue microarray profiling of cancer specimens and cell lines: opportunities and limitations. Lab Invest 81:1331–1338.
4. Fejzo, M.S., D.J. Slamon (2001). Frozen tumor tissue microarray technology for analysis of tumor RNA, DNA, and proteins. Am J Pathol 159:1645–1650.

5. Zhou, L., Hodeib, M., Mendoza, L., Abad, J.D., Z. Hu (2006). New tissue microarray technology for analyses of gene expression in frozen pathological samples. Biotechniques 43:101–104.
6. Simon, R., M. Mirlacher, G. Sauter (2004). Tissue microarrays. Biotechniques 36:98–105.
7. Waschek, J.A., R.A. Casillas, T.B. Nguyen, E.M. DiCicco-Bloom, E.M. Carpenter, W.I. Rodriguez (1998). Neural tube expression of pituitary adenylate cyclase-activating peptide (PACAP) and receptor: potential role in patterning and neurogenesis. Proc Natl Acad Sci 95:9602–9607.
8. Fernebro, E., M. Dictor, P.O. Bendahl, M. Ferno, M. Nilbert (2002). Evaluation of the tissue microarray technique for immunohistochemical analysis in rectal cancer. Arch Pathol Lab Med 126:702–705.

[illegible]

Chapter 10

Building "Tissue" Microarrays from Suspension Cells

Shuchun Zhao and Yasodha Natkunam

Abstract

Tissue microarray (TMA) is a highly efficient method that allows for large-scale measurement of expression of RNA or protein in multiple tissue sections simultaneously. Most TMAs are made from paraffin-embedded tissues. In this chapter, we detail a method that enables construction of TMAs from small volumes of cells in suspension. A TMA is built using pellets of 1×10^6 to 5×10^7 spun cells after fixation, processing, and embedding. The entire procedure is carried out in a microcentrifuge tube and yields excellent preservation of cytomorphology and immunoreactivity from both fresh and frozen suspension cells. It is particularly useful for the study of hematopoietic neoplasms presenting in the blood and bone marrow, fine needle aspirates, and body fluids as well as cultured cells. In addition, this versatile method may facilitate the exploration of gene expression profiling and protein expression in clinical trials where regular tissue biopsies are not available.

Key words: Tissue microarrays, Suspension cells, Fine needle aspirates, Protein expression, Hematopoietic neoplasm

1. Introduction

Tissue microarrays have significantly advanced the study of RNA and protein expression in tumor specimens since it was first reported by Kononen and colleagues (1). This technology allows for the analysis of hundreds of samples concurrently on a single microscope section. Additionally, TMAs can be made from archived tissue specimens and TMA-sections can be preserved for many years such that valuable clinical samples could be reanalyzed when new biomarkers are discovered. Therefore, TMAs are a versatile method for high-throughput analysis of gene and protein expression for diagnostic, prognostic, and therapeutic purposes.

As a comprehensive research tool, TMAs have significant advantages over conventional sections in that they offer improved standardization and reliability since multiple cores from each

Ronald Simon (ed.), *Tissue Microarrays: Methods and Protocols*, Methods in Molecular Biology, vol. 664, DOI 10.1007/978-1-60761-806-5_10, © Springer Science+Business Media, LLC 2010

specimen can be easily incorporated for study. In addition, they are time- and cost-effective as the quantity of antibodies and probes used in the detection of biomarkers on TMAs is minimized. The majority of TMAs, however, are largely dependent on the availability of paraffin-embedded tissues. Material from cell suspensions or needle aspirates have been previously used to construct TMAs; however, these methods require large numbers of cells and excludes the use of small amounts of cell suspensions obtained from clinical samples such as peripheral blood, bone marrow aspirates (BMA), fine needle aspirates, and other body fluids (2–4). Another significant challenge is that some of these previous methods involve embedding cells in an agarose gel matrix, which is then placed in a paraffin mold. Cutting a TMA made of hundreds of agarose cores is technically difficult and requires that numerous agarose-paraffin interfaces be bridged. This technique may cause sample shrinkage or fragmentation. In contrast, we developed a method that allows TMAs to be created from as few as one million cells (5). This method has the added benefit that fixation, processing, and embedding of the cell pellet is carried out using standard histological procedures and yields excellent cytomorphology and immunogenicity. As previously described, pellets of 1×10^6 to 5×10^7 cells are directly fixed, processed, and embedded in paraffin in eppendorf tubes; then a TMA is built by embedding 0.6–2.0 mm cylindrical cores taken from the eppendorf tubes into a recipient paraffin block. These TMAs can then be sectioned routinely. Results from a TMA constructed using cell pellets from 18 BMA and corresponding paraffin-embedded bone marrow core biopsies showed no difference in histologic detail or reactivity to a panel of antibodies (6) (see Fig. 1).

The most important contribution of this technology is that it provides a way to use very small amounts of suspension cells, aspirates, and body fluids in large-scale studies of protein and gene expression for clinical purposes. With the increasing use of small needle core biopsies and fine needle aspirates in diagnostic practice, this technique not only broadens the repertoire of clinical samples used in TMA studies but also facilitates the exploration of gene and protein expression in clinical trials where routine tissue biopsies are unavailable.

2. Materials

2.1. Collection and Fixation of Cultured Cells

1. Cell scraper.
2. 15-mL Centrifuge tubes (BD Falcon, Franklin Lakes, NJ).
3. 0.5-mL, 1.5-mL Microfuge tubes (E&K Scientific, Campbell, CA).
4. Phosphate-buffered saline (PBS), pH 7.2–7.4.

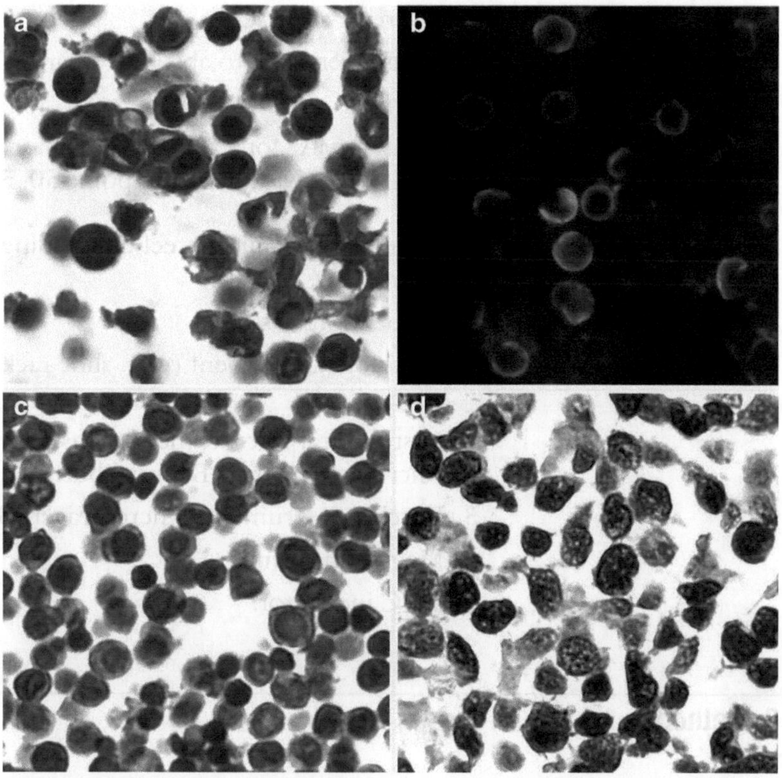

Fig. 1. Cytologic and immunophenotypic results from cell microarrays. Representative examples of H&E staining, immunofluorescence, and immunohistochemistry of microarray cores generated from cell suspensions show excellent cytologic detail and antigenic preservation. H&E staining (**a**), and immunofluorescence staining for CD79a (**b**), of a microarray core with childhood precursor B-acute lymphoblastic leukemia prepared from a bone marrow aspirate; H&E staining (**c**), and immunohistologic staining for nucleophosmin (NPM) (**d**), of an acute myeloid leukemia prepared from a bone marrow aspirate.

5. Microcentrifuge.
6. Benchtop centrifuge.
7. 10% Neutral-buffered formalin.
8. Microscope and counting chamber for determining cell counts.
9. Refrigerator.
10. Shaker.

2.2. Collection, Fixation, Processing, and Embedding of Bone Marrow Aspirates and Cultured Cells

1. RPMI-1640 medium (e.g., Gibco).
2. Fetal calf serum (FCS).
3. RPMI/10% FCS: RPMI 1640 with the addition of 10% FCS (Stable 1 year in the freezer, 2 weeks in the refrigerator).
4. 10-mL centrifuge tubes.
5. Ficoll-HYPAQUE PLUS.
6. Water bath set at 37°C.

7. Oven set at 65°C.
8. Graded series up to absolute ethanol: 70, 95, and 100% (v/v).
9. Xylene.
10. Paraffin wax, for example, Paraplast X-tra (Oxford Labware, St. Louis, MO), (melting point 50–54°C).

2.3. Additional Materials Required for TMA Construction and Sectioning

1. Tissue Tek Stainless-steel embedding molds (Sakura Finetek, Torrance, CA).
2. Embedding cassettes.
3. Histology equipment (e.g., slide racks, slide baths, slides, and cover slips).
4. Histological microtome, for instance, Leica RM2255 (Leica Microsystems Inc., Bannockburn, IL).
5. Beecher Instruments Microarrayer (Silver Springs, MD).
6. Magnifying lamp.
7. 45°C incubator.

3. Methods

All steps are performed at room temperature if not otherwise specified.

3.1. Collection and Fixation of Cultured Cells

1. Collect cells from culture dishes by scraping. Do not trypsinize the cells (see Note 1).
2. Transfer suspended cells into a 15-mL tube.
3. Spin cells at 1,000 × *g* for 5 min at room temperature and remove supernatant.
4. Wash cells with ice-cold PBS.
5. Spin cells as in step 3.
6. Resuspend cells to a density of $1–5 \times 10^7$ cells/mL in ice-cold PBS.
7. Transfer 1-mL cells to xylene resistant 1.5 mL eppendorf tubes.
8. Repeat step 3.
9. Resuspend cells in 1.5 mL 10% Neutral-buffered formalin and fix cells for 1 h with gently rocking.
10. Repeat step 3.
11. Resuspend cells in 1.5 mL PBS.
12. Place cells in refrigerator or start to process (Go to Subheading 3.4 step 1 for processing).

3.2. Collection and Fixation of Fresh Bone Marrow Aspirates

This procedure is applicable for collection and fixation of cells from peripheral blood and BMA; cells of interest (usually white blood cells) can be separated by centrifugation using a Ficoll-Hypaque gradient (7). For other specimens such as fine needle aspirates and body fluids, users can ignore steps 1–7 and start directly from step 8. Cells collected from a variety of sources can be used and the cell collection steps may need appropriate modifications depending on the source of the cells or tissue.

1. Transfer 1 mL fresh BMA to a 15-mL tube containing 3 mL RPMI/10% FCS media, agitate gently.
2. Add 3 mL Ficoll-HYPAQUE PLUS to a 10-mL centrifuge tube.
3. Carefully layer the diluted BMA onto the Ficoll.
4. Centrifuge at 460 × *g* for 30 min at 18–20°C.
5. Remove the upper layer using a 1-mL eppendorf tip, leaving the mononuclear cell layer undisturbed.
6. Using a 1-mL eppendorf tip transfer the mononuclear cells to a clean 10-mL centrifuge tube containing 5 mL RPMI 1640.
7. Agitate the cells gently.
8. Spin at 460 × *g* for 10 min at 18–20°C.
9. Discard the supernatant.
10. Resuspend cells in 0.5 mL PBS.
11. Transfer cells to a xylene resistant 1.5-mL eppendorf tubes.
12. Rinse the centrifuge tube from step 10 with 1.0 mL PBS.
13. Repeat step 11 (see Note 2).
14. Centrifuge at 460 × *g* for 10 min at 18–20°C.
15. Discard the supernatant.
16. Resuspend cells in 1.5 mL 10% Neutral-buffered formalin and fix cells for 1 h with gentle rocking.
17. Repeat step 14.
18. Resuspend cells in 1.5 mL PBS.
19. Place cells in refrigerator or start to process (Go to Subheading 3.4 step 1 for processing).

3.3. Collection and Fixation of Dimethylsulfoxide (DMSO) Preserved Cells

1. Warm water bath to 37°C.
2. Thaw one 1 mL aliquot of RPMI/10% FCS holding media.
3. Remove one BMA vial from freezer and immediately thaw by immersion in the 37°C water bath, agitate gently, and constantly for quick-thaw.
4. Add 100 μL RPMI/10% FCS to BMA, agitate gently for 1 min.

5. Add another 100 μl RPMI/10% FCS, agitate gently for 5 min.
6. Centrifuge at 1,000 × *g* for 5 min and discard supernatant.
7. Resuspend cells in 1.5 mL PBS.
8. Spin cells at 1,000 × *g* for 5 min and remove supernatant.
9. Resuspend cells in 1.5 mL 10% Neutral-buffered formalin and fix cells for 1 h at room temperature with gently rocking.
10. Repeat step 8.
11. Resuspend cells in 1.5 mL PBS.
12. Place cells in refrigerator or start to process (Go to Subheading 3.4 step 1 for processing).

3.4. Processing and Embedding of Fixed Specimen

1. Place fixed cell pellet of 1×10^6 to 5×107 cells in PBS.
2. Centrifuge at 1,000 × *g* for 5 min and remove supernatant.
3. Resuspend cells in 1 mL 70% ethanol by hand mixing.
4. Rock gently for 20 min.
5. Repeat step 2.
6. Resuspend cells in 1 mL 95% ethanol by hand mixing.
7. Rock gently for 20 min.
8. Repeat step 2.
9. Resuspend cells in 1 mL 100% ethanol by hand mixing.
10. Rock gently for 20 min.
11. Repeat step 2.
12. Resuspend cells in 1 mL xylene by hand mixing.
13. Rock gently for 20 min.
14. Centrifuge at 1,000 × *g* for 5 min.
15. Remove supernatant carefully since pellets become translucent after incubation with xylene.
16. Repeat steps 14 and 15 as many times as necessary to remove xylene completely (see Note 3).
17. Incubate cell pellets in 65°C oven for 5 min.
18. Resuspend cells gently in melted paraffin 1–2 times volume of cell pellet (see Note 4).
19. Incubate cells in 65°C oven for a minimum of 20 min with resuspension by tapping the tube gently by hand after the first 10 min.
20. Mix cells gently and thoroughly and place tube immediately in ice water to allow hardening of the cell slurry.
21. Place cells in refrigerator or start to build the TMA.

3.5. TMA Construction and Sectioning

3.5.1. Making a Recipient Paraffin Block

1. Liquefy paraffin pellets in the paraffin dispenser.
2. Place the embedding cassette onto a clean, metal embedding mold.
3. Pour paraffin into the cassette and allow it to cool.
4. Take off the base mold carefully.
5. Cut all irregularities from the edges of the cassette.
6. Face the block until the surface is smooth.
7. Check the block against light; make sure that the block is free of any holes and cracks.

3.5.2. Constructing the TMA

Consider the core size, usually, between 0.6 and 2.0 mm, before TMA construction. Please read Beecher Arrayer Instruction Manual (8). TMA construction is best undertaken under a magnifying lamp.

1. Design a construction key depicting core locations for cases and proper controls (see Note 5).
2. Take a recipient block and label it with the appropriate TMA number.
3. Place the recipient block in the Beecher Arrayer holder firmly.
4. Place the holder with block firmly against the two location curbs on the Beecher Arrayer (the magnets built into the arrayer will hold it firmly in place).
5. Adjust X, Y coordinates for the first hole, leave at least 4 mm distance from top and left edges, and zero the micrometer.
6. Adjust the depth stop properly.
7. Make a hole in the recipient block with the 0.6-mm large punch (see Note 6).
8. Take sample tubes in order based on the construction key.
9. Cut the eppendorf tube at a position 2–3 mm above the paraffin cell pellet.
10. Hold the tip of the tube by hand and take a cylindrical core with the 0.6-mm punch (see Note 7).
11. Place the 0.6-mm cylindrical core in corresponding hole in the recipient block.
12. Adjust *X*, *Y* coordinates for next punch and leave 1.0-mm spacing between the centers of two adjacent cores; repeat steps 7–11 till the TMA is finished.
13. Remove the TMA block from the holder and incubate it for 25 min at 45°C to promote adherence of the cell cores to the walls of the holes in the recipient array block.

14. Place the TMA block on a leveled bench and place a clean glass slide on top of the block. Press the slide gently to make complete contact with the entire surface of the block; level the block surface by using a vial level and leave the slide on top of the block.
15. Put the leveled TMA block with the glass slide in a 4°C refrigerator for a minimum of 30 min to harden the block.
16. Take out the TMA block and remove the slide carefully without tweaking the block surface.
17. Place the TMA block in a microtome and trim the face slightly.

3.5.3. TMA Sections

TMA sections are performed with a standard microtome (for example, Leica RM2255) at 4- to 5-μm thickness. Then sections are floated on a 42°C water bath filled with distilled water until sections flatten. It is important to align the TMA sections before placing them onto slides so that the first core is oriented appropriately to a given end of each slide. Mount sections with Fisher Superfrost Plus glass slides (see Note 8), and place slides vertically on slide-racks and allow them to dry at room temperature. For better preservation of cytomorphology and immunoreactivity of tissues and cells, store slides in a sealed slides container with desiccating agents at 4°C.

4. Notes

1. It is safe to detach cultured cells by gently scrapping since trypsinization might change the protein expression profile. But depending on the cell type and later applications of the cell TMA, cells could be trypsinized without incubation with Trypsin for too long.
2. Take 100 μL of the cell suspension to determine cell concentration by using a counting chamber under a microscope. Make sure to have at least a million cells before fixation. Steps 1–11 can be repeated multiple times to collect more cells.
3. To ensure optimal cell recovery for paraffin embedding, xylene should be removed carefully and completely. This is also critical for building the TMA since xylene residue may prevent cores from being removed due to softening of paraffin.
4. In order to resuspend and distribute the processed cells evenly in paraffin, prewarm the paraffin, pipetors and pipette tips in a 65°C oven for a minimum of 10 min.

5. Determine the numbers of cores for the TMA and leave a minimum of 4 mm distance between the cores and the edges of the recipient block. Samples can be assigned unique numbers (and anonymized if clinical samples) as appropriate. Placing tissue controls (e.g., placenta) in specific locations of the TMA block (e.g., the first two cores) can help orient TMA sections.
6. In our experience, the cores taken from paraffin tubes are softer than the cores taken from a regular paraffin block. It is better to widen the hole by punching twice to facilitate incorporation of the cell-paraffin cores into the recipient block.
7. The depth of the paraffin core for each sample will depend on the cell volume and the total amount of paraffin added to the tube. Repeated punches enable proper core depth for TMAs.
8. We use Superfrost Plus glass slides (Fisher Scientific, Pittsburgh, PA) to mount all TMA sections. These slides electrostatically attract frozen and formalin-fixed tissue sections and enhance the binding of cells and tissues tightly to the slide.

References

1. Kononen J, Bubendorf L, Kallioniemi A, et al. (1998) Tissue microarrays for high-throughput molecular profiling of tumor specimens. Nature Medicine 4:844–847.
2. Andersson A-C, Stromberg S, Backvall H, et al. (2006) Analysis of protein expression in cell microarrays: a tool for antibody-based proteomics. Journal of Histochemistry & Cytochemistry 54:1413–1423.
3. Ferrer B, Bermudo R, Thomson T, Nayach I, Soler M, Sanchez M, Castillo M, et al. (2005) Paraffin-embedded cell line microarray (PECLIMA): development and validation of a high-throughput method for antigen profiling of cell lines. Pathobiology 72:225–232.
4. Moskaluk CA, Stoler MH. (2002) Agarose mold embedding of cultured cells for tissue microarrays. Diagnostic Molecular Pathology 11:234–238.
5. Montgomery K, Zhao S, van de Rijn M, Natkunam Y. (2005) A novel method for making "tissue" microarrays from small numbers of suspension cells. Applied Immunohistochemistry and Molecular Morphology 13:80–84.
6. Hazard F, Zhao S, Schiffman J, Lacayo N, Dahl G, Natkunam Y. (2008) Tissue microarrays from bone marrow aspirates for high throughput assessment of immunohistologic markers in pediatric acute leukemia. Pediatric and Developmental Pathology 11(4):283–290.
7. Boyum, A. (1968) Separation of leukocytes from blood and bone marrow. Scandinavian Journal of Clinical and Laboratory Investigation 21:77.
8. Instruction manual of Manual Tissue Arrayer: Beecher Instruments, Inc., Sun Prairie, WI.

Chapter 11

Tissue Microarrays from Biopsy Specimens

Milton W. Datta and André A. Kajdacsy-Balla

Abstract

The construction of tissue microarrays from needle core biopsy specimens allows for the study of diseases with limited tissue samples and provides insight into tumors that are treated with nonsurgical approaches. While the techniques are technically challenging, in the hands of investigators they can reveal new insights into biomarkers of disease.

Key words: Tissue microarray, Biopsy, Pathology, Prostate cancer

1. Introduction

1.1. Needle Core Biopsies in Pathologic Diagnosis and Biopsy Tissues

Biopsies have been used for centuries in the diagnosis of human pathology and have undergone a rapid increase in use with the advent of endoscopic and laparoscopic procedures (1). While sufficient tissue is often obtained for pathologic diagnosis and clinical treatment, only a small fraction of material is left for further study. As particular disease processes and treatments are now being diagnosed and treated by these minimally invasive procedures, the study of certain diseases and their treatments will depend upon research studies conducted on the remnants of these very small samples. Examples would include needle core biopsies of tumors subsequently treated by radiation or chemotherapy, without prior surgical excision.

1.2. Methods for Production of TMAs from Needle Core Biopsies

The rapid evaluation of new biomarkers for clinical and prognostic relevance has led to the use of tissue microarrays, yet the use of biomarkers has often been limited to larger tissue resections, and has not included small needle core biopsies. More recently, two methods for the construction of tissue microarrays from prostate needle core biopsy specimens have been described (2, 3).

Ronald Simon (ed.), *Tissue Microarrays: Methods and Protocols*, Methods in Molecular Biology, vol. 664,
DOI 10.1007/978-1-60761-806-5_11, © Springer Science+Business Media, LLC 2010

Similar methods have been used on larger bone marrow core biopsy specimens (4). One of the two main methods is described here, and while it is technically challenging, it provides a method for the effective use of small tissue samples for tissue microarray production and subsequent high throughput biomarker evaluation.

1.3. Recent Publications Using Needle Core TMAs

The described methods for the production of tissue microarrays from prostate needle core biopsies and our ability to evaluate immunohistochemical biomarkers on these slides have been described in detail in the original publications (2, 3). In our hands, over 100 slides can be made from a biopsy tissue microarray, and this can yield accurate biomarker data (2). Furthermore, biopsy tissue microarray slides have been used in successful clinical studies to evaluate Ki-67, VEGF, and HIF-1 alpha in prostate cancer biopsies (5, 6). An additional study has described a method similar to Jhavar et al. that was used for the construction of a tissue microarray from bone marrow biopsies (4). One publication has described disappointing results from a tissue microarray produced from prostate needle biopsies, but examination of their publication demonstrates a failure to follow either of the methods outlined in the prior publications (7). Thus we would recommend at least trying to follow the methods described here to enhance one's chances of success.

1.4. The Future Use of Needle Core TMAs

Certain diseases or tumors will continue to be treated aggressively after initial diagnosis on small needle biopsy specimens. In these situations, where subsequent radiation or chemotherapy may be used, these small initial biopsies may be the only material available for molecular or biomarker studies. The ability to produce tissue microarrays from these small biopsy samples will be important in the study and evaluation of biomarkers for better diagnosis, prognosis, and treatment outcome stratification.

2. Materials

1. MTA1 TMA arrayer (Beecher Instruments, Sun Prairie WI).
2. 1.5 mm corer (Beecher Instruments, Sun Prairie WI).
3. Paraffin blocks with plastic cassettes.
4. Large rubber tube, (catalog #1056-44, Fernco).
5. Cutting knife.
6. Dremel tool Model #770, (Dremel Tools, Racine WI).
7. 1.25 in. cutoff wheel #540 (Dremel Tools, Racine WI).
8. Heavy Duty Aluminum foil (catalog # 01-213-2, Fisher Scientific, Pittsburgh, PA).

9. Glass capillary tubes, (Cat #1909339, Roche-Lightcycler, Indianapolis IN).
10. Drying oven, 70°C.
11. Melted paraffin (Cat# 8337, Richard-Allan Scientific, Kalamazoo, MI).
12. Shallow pan with ice water.
13. Pipetter and pipette tips.
14. Single edge razor blades.

3. Methods

While both Datta et al. and Jhavar et al. have been successful in the production of tissue microarrays from needle core biopsy specimens, the methods differ in their technical methodology, material, and time requirements. The method used by Jhavar is simple and inexpensive, but there has also been some difficulty in reproducibility and sample representation on the subsequent TMA slides (7). Datta's method is technically more complex, but has been successful in the production of larger numbers of biopsy TMA slides for projects. The latter is presented here for use by the reader.

3.1. Planning and Preproduction of Needle Biopsy TMAs

3.1.1. Sample Numbers and Tissue Representation

Due to the small sample sizes in biopsy specimens, there is a need for careful sample size estimations based on tissue loss and sampling bias (2). While a full discussion of sample size analysis and TMA design are beyond the scope of this chapter, one is referred to the previous chapters by Sauter, Mirlacher, and Simon. Detailed studies of sample size and molecular characterizations have been published for prostate cancer biomarkers (8) and have been reviewed (9) and are often used as a starting point for TMA production in other systems. Previous work by the authors has suggested general criteria for biopsy TMA production and is summarized as follows:

- The length of the biopsy with tumor/diagnostic tissue affects the number of biopsy TMA slides that can be produced, with an optimal length of ≥2.0 mm.
- The number of biopsy tumor/diagnostic tissue areas that can be placed on the biopsy TMA affects accurate specimen representation, with a minimum of three and an ideal of five representative portions needed for >90% specimen representation over 100 biopsy TMA slides.

3.1.2. TMA Design Layout Considerations Specific for Needle Core TMAs

Interspersed within the needle core tissues, place a series of large tissue orienting cores. We prefer to use liver tissue cores from autopsy cases (see Note 1).

3.1.3. Specimen Identification, Pathologic Verification, Slide and Block Marking

1. Confirm the presence of the tissue of interest on the H&E stained slide and corresponding tissue block (see Note 2).
2. Mark the corresponding portion of the paraffin tissue block that corresponds to the area of interest in the H&E stained slide using a marking pen (Sharpie, Sanford Corp., Oak Brook IL, USA) or simply a metal tip to score the surface of the paraffin block. Once marked on the biopsy paraffin blocks, the areas of interest for subsequent biopsy TMA production are easily identified for sampling (Fig. 1a).

3.2. TMA Block Production

The subsequent steps are presented in Fig. 1, reprinted with permission from Datta, 2005.

3.2.1. Method of Needle Core Biopsy TMA Production, Preparation of the Recipient TMA Block

1. Create a donor TMA block using the Beecher system or equivalent with a 1.5 mm corer.
2. Place the orienting liver tissue cores in the donor TMA block according to the planned specimen grid design (Fig. 1k).

3.2.2. Preparation of the Biopsy Template and Foils

1. A biopsy trough is cut into a flat piece of rubber tube ($6 \times 0.7 \times 0.5$ in.) using the Dremel tool and 1.25-in. cutoff wheel (Fig. 1b, c). The tool is used to cut 3 mm deep troughs into the rubber block at 1.5 cm intervals. The groove should be no wider than the cutting blade, as wider troughs will require subsequent trimming of the biopsy paraffin cores and additional work.
2. Each trough in the rubber tube is lined by 0.7×0.7 in. pieces of aluminum foil. The foil is prepared by wrapping around the length of a capillary tube and then pressing the foil/tube into the trough of the rubber tube (Fig. 1d). The foil ends are then folded flat to the surface of the rubber tube creating a butterfly shape that lines the trough and adjacent rubber surfaces.

3.2.3. Creation of Biopsy Specimen Cores

1. The specimen samples corresponding to the marked areas of the biopsy paraffin block are wedged out of the block using a single edge razor blade, and placed in one of the foil-lined shallow troughs (Fig. 1e, f). We often mark the location of the tissue and other notes on the foil for sample tracking (see Note 3).
2. Place the rubber tube with the foil troughs and samples in a drying oven for 5 min to melt the paraffin.

3. The rubber tube with foiled specimens is then removed from the oven and the remainder of the trough is then filled with melted paraffin using 200 μl pipette tips (Fig. 1g).
4. After filling the trough, the rubber tube and foiled specimens are placed in the ice water bath for 2 min.
5. The foils and specimens are removed from the rubber tube and trimmed with three cuts of a single edge blade, one to trim excess paraffin off the top of the trough, and two cuts on either end of the sample to create a cylinder that can be placed vertically in the recipient TMA block (Fig. 1h, i). The cylinder is prepared such that tissue is present at one end (the end

Fig. 1. Preparation and use of the biopsy tissue microarray template. (**a**) Color marking of biopsy tissue specimens. Using the corresponding H&E stained slide, areas of prostate cancer are marked with an ink pen. (**b**) Cutting of the rubber biopsy template block. (**c**) Preparing biopsy troughs using a Dremel tool. (**d**) Preparation of foil inserts for biopsy samples. (**e**) Wedge resection of marked biopsy tissue. (**f**) Placement of the wedged tissue core into the foil template, the tissue side is facing down.

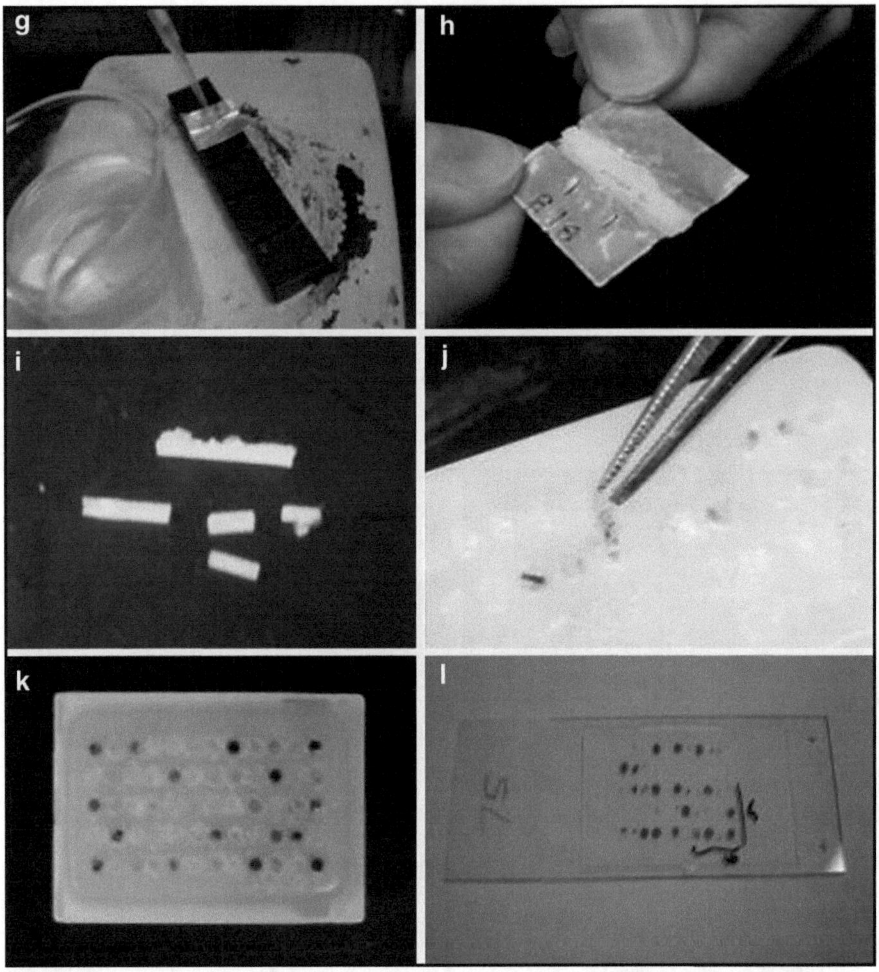

Fig. 1. (continued) (**g**) Injection of additional paraffin into biopsy template troughs. (**h**) Opening of biopsy foil trough to reveal the paraffin wedge with biopsy tissue. One can see the tissue in the paraffin wedge due to the ink staining. (**i**) Size cutting of the paraffin wedge with biopsy tissue to fit the microarray recipient block hole. Note the waste paraffin core (foreground) from recipient donor block production being used for size standardization. (**j**) Placement of the biopsy core into the microarray recipient block. (**k**, **l**) Completed biopsy tissue microarray block and hematoxylin and eosin stained slide. Note the orienting liver tissue cores. Reprinted with permission from ref. (2).

on the surface of the recipient paraffin block) so that the initial cuts from the finished TMA block will provide tissue on slides.

6. Using tweezers, each core is placed in the recipient block and aligned with the block surface (Fig. 1j) (see Note 4).

3.2.4. Slide Preparation from TMA Blocks

1. The finished TMA paraffin block is faced and slides prepared in the usual fashion (5 μm sections floated on warm water and placed on Superfrost Plus positively charged slides) (see Note 5).
2. The initial slides are stained by H&E for biopsy evaluation and quality assurance. The remaining slides are then used for

research studies. We recommend that every tenth slide be H&E-stained for quality assurance purposes to evaluate for the presence and histologic localization of the disease process of interest.

4. Notes

1. Because of the small nature of the specimens, there is a need for accurate methods for tissue and slide orientation. This becomes critical upon slide preparation, where tissue folding, partial sectioning or tissue stretching artifacts can distort the grid of the array and confuse sample scoring. For this purpose, we have found that the placement of traditional large tissue cores of an easily identifiable sample can greatly facilitate sample orientation and allow for more accurate biopsy TMA scoring. We recommend the use of strategically inserted large orienting tissue cores in the biopsy TMA donor block. Liver tissue cores, which can be easily obtained from autopsy cases, can be useful for this purpose.
2. Key to the success of both methods is the accurate identification and marking of the tumor or lesion in the biopsy paraffin block. This process starts with the accurate pathologic review of the initial biopsy H&E stained slides, preferably by a trained pathologist. We prefer to use the initial diagnostic slide for the purposes of biopsy review, as this avoids the need for additional sectioning of the biopsy paraffin block, a process that needs block facing and recutting to obtain a flat surface and subsequent tissue levels produced for H&E staining. This unnecessary step can easily deplete both the diagnostic tumor and in some cases the majority of the remaining biopsy specimen, leaving insufficient material for tissue microarray production. Upon reviewing the slide the pathologist should mark the sections of the biopsy specimen core that represent diagnostic tumor and control normal tissue. The marked slides are then correlated with the corresponding paraffin block and additional lines are placed on the paraffin block denoting the tissue of interest by the pathologist (Fig. 1b).
3. The small size of biopsy specimens makes the production of tissue microarrays from the remaining tissue technically challenging. The excision of the remaining tissue from the biopsy paraffin block must be handled carefully, as limited material requires very careful removal. For paraffin embedded biopsy samples where multiple slides were prepared for diagnostic purposes, very limited material may be left for TMA production. In these situations, batching of the remaining tissue

fragments may be the best way to maximize the material for TMA production. We have accomplished this by removing multiple tissue fragments from the donor paraffin block and melting them together in the drying oven in Subheading 3.2.3.

4. If a biopsy tissue cylinder is too small for the core hole, the sample can be rapidly dipped in melted paraffin to increase the width of the cylinder before placement in the recipient block.
5. Considering how difficult it is to obtain some of the specimens, it is usually recommended to use a collimator (Histo Collimator, Model HM-355, Microm International, Waldorff, Germany) when facing the TMA blocks in order to avoid loss of significant portion of cores on one side of the TMA. Because it inevitable to lose some portions of the donor TMA block every time one has to face the block, it is recommended to cut all the slides in a single sitting or at least cut a batch that will be used within a few days. The drawback of cutting slides that will not be immediately used is the loss of antigenicity for some markers, probably due to prolonged exposure to air (10). In order to avoid oxidation with time, it is recommended that slides that will not be used right away should be stored in nitrogen gas chambers after thick paraffin recoating of sections (10). While some groups have used tape transfer for TMA slide production, we have not found this to be necessary for prostate needle biopsy TMAs.

Acknowledgments

This work has been supported in part by a grant from the National Cancer Institute (U01 CA 86473). We would like to thank the members of the Cooperative Prostate Cancer Tissue Resource, whose support and enthusiasm made the development of these techniques possible.

References

1. Zerbino, D.D. (1994) Biopsy: its history, current and future outlook. *Lik Sprava.* **3–4**:1–9.
2. Datta, M.W., Kahler, A., Macias, V., Brodzeller, T., Kajdacsy-Balla, A.A. (2005) A simple inexpensive method for the production of tissue microarrays from needle biopsy specimens: examples with prostate cancer. *Appl Immunohistochem Mol Morphol.* **13**(1): 96–103.
3. Jhavar, S., Corbishley, C.M., Dearnaley, D., Fisher, C., Falconer, A., Parker, C., Eeles, R., Cooper, C.S. (2005) Construction of tissue microarrays from prostate needle biopsy specimens. *Br J Cancer.* **93**(4):478–82.
4. Obermann, E.C., Marienhagen, J., Stoehr, R., Wuensch, P.H., Hofstaedter, F. (2005) Tissue microarray construction from bone marrow biopsies. *Biotechniques.* **39**(6):822, 824, 826.

5. Jhavar, S., Bartlett, J., Kovacs, G., Corbishley, C., Dearnaley, D., Eeles, R., Khoo, V., Huddart, R., Horwich, A., Thompson, A., Norman, A., Brewer, D., Cooper, C.S., Parker, C. (2009) Biopsy tissue microarray study of Ki-67 expression in untreated, localized prostate cancer managed by active surveillance. *Prostate Cancer Prostatic Dis.* **12**(2):143–7.
6. Vergis, R., Corbishley, C.M., Norman, A.R., Bartlett, J., Jhavar, S., Borre, M., Heeboll, S., Horwich, A., Huddart, R., Khoo, V., Eeles, R., Cooper, C., Sydes, M., Dearnaley, D., Parker, C. (2008) Intrinsic markers of tumour hypoxia and angiogenesis in localised prostate cancer and outcome of radical treatment: a retrospective analysis of two randomised radiotherapy trials and one surgical cohort study. *Lancet Oncol.* **9**(4):342–51. Epub 2008 Mar 17.
7. Singh, S.S., Mehedint, D.C., Ford, O.H. 3rd, Maygarden, S.J., Ruiz, B., Mohler, J.L. (2007) Feasibility of constructing tissue microarrays from diagnostic prostate biopsies. *Prostate.* **67**(10):1011–8.
8. Rubin, M.A., Dunn, R., Strawderman, M., Pienta, K.J. (2002) Tissue microarray sampling strategy for prostate cancer biomarker analysis. *Am J Surg Pathol.* **26**(3):312–9.
9. Kajdacsy-Balla, A.A., Geynisman, J.M., Macias, V., Setty, S., Nanaji, N.M., Berman, J.J., Dobbin, K., Melamed, J., Kong, X., Bosland, M., Orenstein, J., Bayerl, J., Becich, M.J., Dhir, R., Datta, M.W., Cooperative Prostate Cancer Tissue Resource. (2007) Practical aspects of planning, building, and interpreting tissue microarrays: the Cooperative Prostate Cancer Tissue Resource experience. *J Mol Histol.* **38**(2):113–21.
10. DiVito, K.A., Charette, L.A., Rimm, D.L., Camp, R.L. (2004) Long term preservation of antigenicity in tissue microarrays. *Lab Invest.* **84**:1071–8.

Chapter 12

Immunohistochemical Analysis of Tissue Microarrays

Ronald Simon, Martina Mirlacher, and Guido Sauter

Abstract

Immunohistochemistry (IHC) is the gold standard methodology for in-situ protein expression analysis in tissue samples. The combination of IHC and tissue microarray (TMA) technology allows for the simultaneous analysis of hundreds of tissue samples with an unprecedented degree of experimental standardization. The same immunostaining protocols used for conventional large sections can be used for TMAs, including antigen retrieval procedures for staining of routinely archived formalin-fixed tissue samples. The development of optimal IHC protocols is highly important for TMA studies because minor protocol variations often have a marked impact on the outcome of the staining. Preabsorption and isotype-specific control experiments should be included as the last step in protocol development to proof target protein-specific binding. Such controls are particularly important for new antibodies with unknown staining patterns.

Key words: Immunohistochemistry, Formalin fixed tissue samples, Antibody protocol development, Antigen retrieval, IHC controls

1. Introduction

The vast majority of published studies using tissue microarrays (TMAs) have employed immunohistochemistry, which is the gold standard method for in-situ protein expression analysis in tissue samples. In this technique, antibodies are used to visualize proteins directly in their natural cellular localization. In a typical IHC experiment, an unlabeled primary antibody, which is specific for the "target" protein of interest, is combined with a secondary labeled antibody directed against the primary antibody. Multiple systems for further signal amplification are on the market. Two classes of labels are commonly used to detect the secondary antibody. Fluorochromes like FITC or rhodamine or its derivates are used if simultaneous visualization of two or more target proteins is required.

Ronald Simon (ed.), *Tissue Microarrays: Methods and Protocols*, Methods in Molecular Biology, vol. 664,
DOI 10.1007/978-1-60761-806-5_12, © Springer Science+Business Media, LLC 2010

For standard IHC experiments detecting one protein at a time, for example, in routine pathological diagnosis, enzymatic labels reacting with brightfield dyes like 3.3-Diaminobenzidine (DAB) or Alkaline Phosphatase-Anti-Alkaline Phosphatase Complex (APAAP) are usually preferred because the staining remains stable for years, and no special equipment is required for microscopy.

The same IHC protocols can be used for tissue microarray (TMA) sections as for conventional large sections. Considering the significant workload connected to TMA analysis, all efforts should be made to perform the analysis with the highest possible quality. It is important to note that the success of such a study mainly depends on the quality of the immunostaining experiment, rather than on the skills of the interpreter or the statistician. A non-suitable antibody causing high levels of background staining or showing poor specificity will at least massively complicate – and most likely adulterate – the entire study. Even minor protocol changes like a suboptimal antibody dilution may obscure relevant findings, for example, if weak expression is missed in a significant fraction of samples because of over-dilution, or if faint background staining is over-interpreted as being positive in case of overstaining. Although such mistakes will become obvious in some studies when the results of known markers do not match the expectations, in many cases – particularly with new markers – there is a high risk of missing important findings if a suboptimal protocol is employed. For these reasons, it is of utmost importance to define the optimal IHC protocol before the study is conducted.

1.1. Antigen Retrieval and IHC Protocol Optimization

For immunohistochemistry of formalin-fixed paraffin-embedded (FFPE) tissues, it is almost always necessary to pre-treat the tissues in order to reverse the fixation induced cross-linking of proteins that typically obscures the epitopes targeted by the antibodies. The most commonly used antigen-retrieval strategy includes treatment of the tissue sections with high temperature in buffers providing variable pH (Heat-Induced Epitope Retrieval, HIER) (1). A properly conducted HIER can dramatically improve the intensity of immunostaining (Fig. 1). Unfortunately, it is not possible to predict which conditions might be best suited for a particular antibody. Some antibody manufacturers give recommendations in their datasheets that may be used as a starting point for protocol optimization, and it may also be helpful to search for published protocols in the literature. However, in many cases such “3rd party” protocols will not yield satisfactory results in the own laboratory. The main reason for this is that there is no special standardized equipment available for IHC from commercial sources. Laboratory devices commonly used for slide pretreatment, including water baths, pressure cookers, microwave ovens, steamers, or autoclaves, are available from a multitude

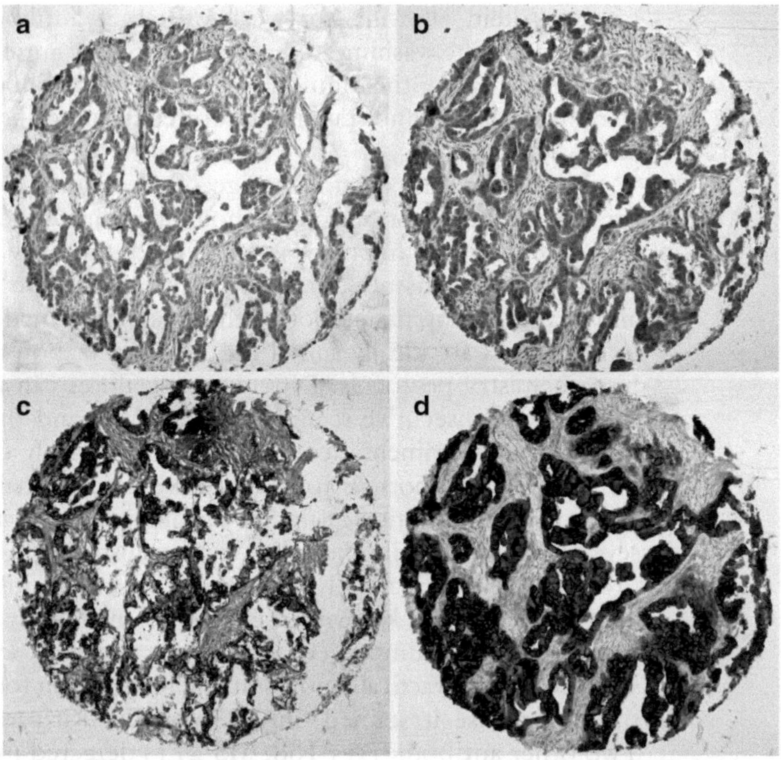

Fig. 1. Effect of different antigen-retrieval strategies on immunostaining quality in an ovarian cancer tissue spot in consecutive sections of a TMA. (**a**) Autoclave pretreatment in citrate buffer pH 6. (**b**) Microwave pretreatment in citrate buffer pH 6, (**c**) 15-min pronase solution. (**d**) Autoclave pretreatment target retrieval solution pH 2.

of manufacturers and show significant differences in power and performance. For example, IHC protocols often state "autoclave pretreatment," i.e., incubation at 120°C for 5 min. However, the time period for heating and cooling of autoclaves from different manufacturers may vary significantly. Some older devices may be opened almost immediately after venting while others need to cool down to room temperature before the door can be unlocked. Consequently, IHC protocols for FFPE tissues (including large sections and TMAs) cannot be easily transferred from one lab to another, but must be adjusted according to the available equipment. No antigen retrieval is required for immunostaining of frozen tissues, where the proteins are usually conserved in their native conformation. If an antibody turns out to be unsuited for staining of FFPE tissues, it may be worth trying frozen tissues instead.

Optimal IHC conditions should result in a crisp staining without any, or only minimal, background staining. Background staining typically occurs in the cell's cytoplasm, and is caused by diffuse, non-specific binding of the antibody. The cytoplasm is a dense mixture of all kinds of proteins including protein degradation products, some of which may exert sufficient similarity to the target protein allowing for weak off-target antibody binding. Extension of the washing steps after the primary antibody incubation, or increasing the dilution of the primary antibody may help to reduce background staining, but often it cannot be completely removed.

For an optimal dynamic range of the immunostaining, the primary antibody concentration should be adjusted so that a wide range of staining intensity is observed, optimally with some strongly positive tissue spots, some entirely negative tissue spots, and the remaining tissue spots with variable intermediate staining levels. TMAs are ideally suited for the optimization of immunohistochemistry protocols. Different procedures can be tested in parallel on consecutive sections of the TMA, and the impact of the various experimental conditions can be quickly compared in the small TMA spots. In one study, we used a small test TMA composed from 80 tissue samples to compare the performance of seven different antibodies directed against the KIT receptor (Table 1) (2). Remarkably, only three of seven antibodies were able to confirm the known Kit expression in all arrayed gastrointestinal stroma tumors (GIST). One antibody (Santa Cruz sc-13508) was practically non-suitable for IHC on formalin-fixed tissues because it stained only 13% of the Kit-positive GIST. Two other antibodies (cs-168, MS-271) detected membranous staining in all GIST but showed intense non-specific cytoplasmic co-staining in virtually all analyzed tissues.

1.2. Controls

The last step of protocol development must include a control experiment testing for non-specific staining, which is a common

Table 1
Comparison of seven different antibodies directed against the KIT protein

Antibody	% GIST[a] (*n*=8)	% Other tumors[a] (*n*=72)	Background staining in stromal cells
A4502	100	54	–
Nr. 566	88	40	–
NCL-CD117	50	42	–
sc-13508	13	3	–
sc-1494	88	71	++
MS-271	100	97	++
sc-168	100	100	++

GIST, gastrointestinal stroma tumor
[a]Positive tumors. With permission from Tapia et al. (5)

problem in IHC experiments, particularly if non-established, poorly characterized antibodies are used. Such antibodies may be directed against a different protein as expected, or may simply "stick" non-specifically to the tissue typically with the "tail," i.e., the c-fragment (Fc) of the heavy chain. Polyclonal antibodies are generally more prone to non-specific binding than monoclonal antibodies because the risk of cross-reactivity with proteins with similarity to the target protein is higher in the polyclonal antibody mixture. However, there is no general rule that mono- or polyclonal antibodies are better suited for IHC, and the choice of antibody should firstly depend on the quality of the staining. Non-specific binding should be considered, for instance, in the case of unusually high-level background staining, if staining occurs in virtually all cell types or with an unexpected high frequency, or if the observed staining pattern does not reflect the anticipated intracellular localization of the target protein. For instance, IHC of a transmembrane receptor should result in membranous staining, whereas a DNA-binding transcription factor can be expected to yield a nuclear staining pattern, although this rule should be applied with caution because many proteins shuttle between the cellular components, and not all functions of a given protein may be known to date.

There are three types of controls that should be performed before a newly established antibody and IHC protocol is used for large-scale studies. A preabsorption control experiment reveals specific binding to proteins other than the designated target protein, an isotype control experiment identifies non-specific binding to any cellular structures, and the primary antibody may be omitted in order to find staining induced by the detection system (e.g., the secondary antibody) alone.

To perform a pre-absorption control experiment, the target protein must be available. Many antibody manufacturers also provide the amino acid sequence of the peptide used for immunization of the host animal in the antibody datasheet. This information can be used to obtain a commercially synthesized peptide. For the blocking experiment, the primary antibody and excess (50-fold) amounts of the target protein/peptide are premixed before they are applied to the tissue section. The pre-absorption control experiment is performed in parallel to the regular IHC experiment to allow for a direct comparison of the results. A negative or at least dramatically reduced staining in the control experiment suggests that the staining seen in the regular experiment is target specific, because the blocking protein completely absorbed the primary antibody before it could bind to the tissue. Any staining found in the blocking experiment as well as in the regular experiment must be considered non-specific for the target protein ("off-target" binding). The pre-absorption control is particularly important for polyclonal antibodies, where some clones might bind specifically and some non-specifically (Fig. 2).

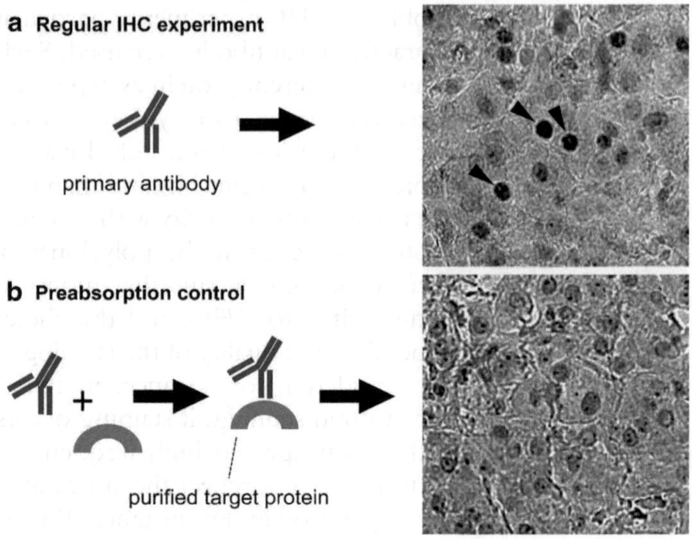

Fig. 2. Preabsorption control. (**a**) In a regular IHC experiment, the antibody binds to the target tissue (cell nuclei, arrowheads). (**b**) In the preabsorption control experiment, the primary antibody is premixed with the purified target protein before it is applied to the tissue. The purified protein specifically binds the antibody so that the IHC experiment appears to be negative.

Isotype specific controls are only applicable for monoclonal primary antibodies. An isotype-specific control antibody is identical to the primary antibody with respect to the immunoglobulin subtype ("isotype") but is directed against an exogeneous, non-human target. As a consequence, isotype control antibodies cannot specifically bind to any component of human tissues, and all staining found with an isotype control antibody must be attributable to non-specific binding (Fig. 3). Such binding is typically mediated by the Fc-portion (i.e., the "tail") of the antibody. The control experiment is carried out in parallel to the regular experiment with the target-specific antibody under exactly the same experimental conditions including antibody concentration and antigen-retrieval procedure.

Omission of the primary antibody highlights non-specific staining that is induced by the detection system only. This kind of control is particularly important if non-standard detection systems are employed, or if major modifications are made to established detection systems. The commonly used commercial detection kits including, for example, the EnVision system (DAKO, Glostrup, Denmark) or the Vectastain ABC-system (Vector Laboratories, Inc. Burlingame, CA, USA) are virtually background-free. In many studies, omission of the primary antibody is the only control performed. Especially in the case of modern detection systems, this control has the least relevance and is not sufficient – if performed alone – to suggest specificity of antibody binding.

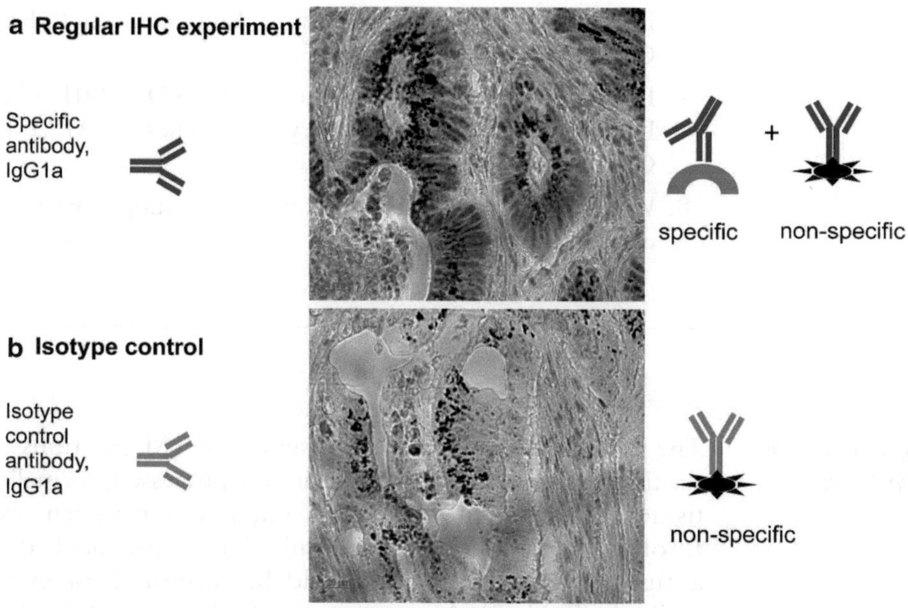

Fig. 3. Isotype specific control. (**a**) In a regular IHC experiment, the antibody binds to both the nuclei of colon epithelium and to the apical pole of the cells. (**b**) In the isotype control experiment, the control antibody binds to the apical cell pole, indicating non-specific binding of IgG1a antibodies. Only the nuclear staining observed in the regular experiment is specific.

2. Materials

2.1. Antigen Retrieval (Formalin Fixed Tissues)

1. Xylene
2. Autoclave
3. Steamer
4. Microwave oven
5. Pressure cooker
6. Pretreatment buffers pH 2, pH 6, pH 8, pH 10, for instance, "Retrievit Target Retrieval Solutions" 10× conc. 1/10 in H_2O (Inno Genex, San Ramon, CA, USA)
7. 15–20 sections of a positive control tissue with known expression of the target protein (conventional large sections or sections of a test TMA)

2.2. Immunohistochemistry

1. Primary antibody
2. Antibody diluent (DAKO, Glostrup, Denmark)
3. Blocking reagent: 1% H_2O_2 in methanol
4. TBS wash buffer: (DAKO, Glostrup, Denmark)
5. Hematoxylin solution (Medite, Burgdorf, Germany)

6. Eukitt slide mounting medium (Kindler GmbH, Freiburg, Germany)
7. EnVision detection kit for mouse (DAKO #4001, Glostrup, Denmark) or rabbit primary antibodies (DAKO #4003, Glostrup, Denmark)
8. Wet chamber, or a plastic box with lid equipped with a moistened paper towel

3. Methods

3.1. Antigen Retrieval (Formalin-Fixed Tissues)

The procedure described here is suited for TMA sections or conventional large sections from routinely processed, formalin-fixed tissues. No antigen retrieval is required for frozen sections. In order to establish the best-suited IHC protocol, different antigen-retrieval strategies should be compared for every new antibody (Table 2). An extensive antibody protocol development strategy may include parallel testing of multiple different pretreatment conditions in positive and negative control tissues (see Note 1).

1. Incubate tissue sections or TMA slides for 1 h in xylene at room temperature to remove paraffin (in fume hood).
2. Rinse slides in ascending series of ethanol: 100%, 95%, 80%, 70%, 50% for 2 min each.
3. Rinse slides in ddH_2O.
4. Place slides in a coplin jar with pretreatment buffer (try buffers of different pH in parallel for protocol development) and perform antigen retrieval according to Table 2. Try different procedures in parallel for protocol development.
5. Transfer slides in a coplin jar with TBS buffer for 5 min and proceed with immunostaining.

3.2. Immunostaining

The following protocol is suitable for detection of mouse or rabbit primary antibodies in formalin-fixed tissues. If primary antibodies from other species are used, an appropriate horseradish peroxidase (HRP) conjugated secondary antibody may be used instead of the DAKO Labeled Polymer (see Note 2). The protocol can also be used for frozen tissues if the large section or TMA section is fixed in cold (4°C) acetone for 10 min and air-dryed.

1. Incubate tissue sections or TMA slides for 10 min in a coplin jar with blocking solution at room temperature in order to block endogeneous peroxidase (see Note 3).
2. Rinse slides in TBS buffer for 5 min, repeat step once.
3. Wipe off the back of each slide and around the tissue but do not touch the tissue. The tissue must be kept wet at all times.

Table 2
Commonly used procedures for antigen retrieval

Device	Temperature	Pressure	Incubation time	Buffer/enzyme	Remarks
Pressure cooker	140°C	Approx. 370 kPa	5–10 min	pH 2-pH 10	Actual temperature and total incubation time (including heating and cooling) largely depends on device
Autoclave	120°C	Approx. 103 kPa	5–10 min		
Microwave	100°C		10 min		
Steamer	~95°C		10 min		Actual temperature depends on device
Coplin jar	Room temperature (~21°C)		15 min	Pronase solution	

4. Place slides in a wet chamber and add 100–150 μl of prediluted primary antibody (in antibody diluent) to each slide (for protocol development try different concentrations; use 10 μg/ml as a starting point).
5. Close the wet chamber and incubate slides for 2 h at 30°C.
6. Rinse slides in TBS buffer for 5 min, repeat step once.
7. Add 1–3 drops of Labeled Polymer (comes with DAKO kit) to completely cover the tissue. Incubate in the wet chamber at room temperature for 30 min.
8. In the meantime, prepare DAB solution: Add 1 drop 3b (liquid DAB and chromagen in DAKO kit) to 1ml solution 3a (Buffered Substrate in DAKO kit). Prepare more solution if necessary (150 μl is needed per slide). Leave the mixed solution at room temperature for 30–60 min before using.
9. Rinse slides in TBS buffer for 5 min, repeat step once.
10. Rinse slides in ddH_2O for 2 min.
11. Wipe off the back of each slide and around the tissue but do not touch the tissue. The tissue must be kept wet at all times.
12. Add 150 μl DAB solution to each slide and incubate at room temperature for 10 min.
13. Rinse slides in ddH_2O for 1 min.
14. Incubate in 1% hematoxylin solution for 1 min at room temperature.
15. Rinse in tap water for 1 min.
16. Dehydrate in ascending series of ethanol (80%, 96%, 3 × 100%, for 2 min each).
17. Dip slides into a coplin jar with xylene.
18. Add 1–2 drops of Eukitt mounting medium and put on a coverslip, removing all air bubbles, and let dry.

3.3. TMA Analysis

All slides of one TMA study are usually incubated in one set of reagents in order to assure identical experimental conditions across all arrayed samples. Due to this unprecedented standardization, surprising variations can occur if experiments are repeated under slightly different conditions. Often, these variations can even alter the threshold for detection of positivity, thus affecting the overall frequency of positive cases. In contrast, associations between examined parameters and clinico-pathological data are usually unchanged, because all groups within one TMA (low and high stage, good and bad prognosis) are equally affected by experimental variations. Large numbers of samples on a TMA, generally increase the likelihood of finding significant associations, especially in case of suboptimal IHC.

For a standard microscopic tumor analysis, the percentage of positive cells, the staining intensity (0, 1+, 2+, 3+), the intracellular localization (membraneous, cytoplasmatic, nuclear), and tissue localization of the staining (tumor cells, stroma, vessels) is recorded. These data can be entered into an ideogram of the TMA during microscopy, and later transferred into a spreadsheet application, for example, Excel (Fig. 4, see Note 4).

For statistical analyses, tumors can be classified into three or four groups based on the fraction of positive cells and the staining intensity. For practical purposes, it is important to have a group with zero staining and one with intense staining in a high proportion of cells. If a biomarker has clinical significance, statistical differences will always be seen between these groups. Further differentiation of intermediate staining results does not have a very high impact on the study results. Generally, the intermediate

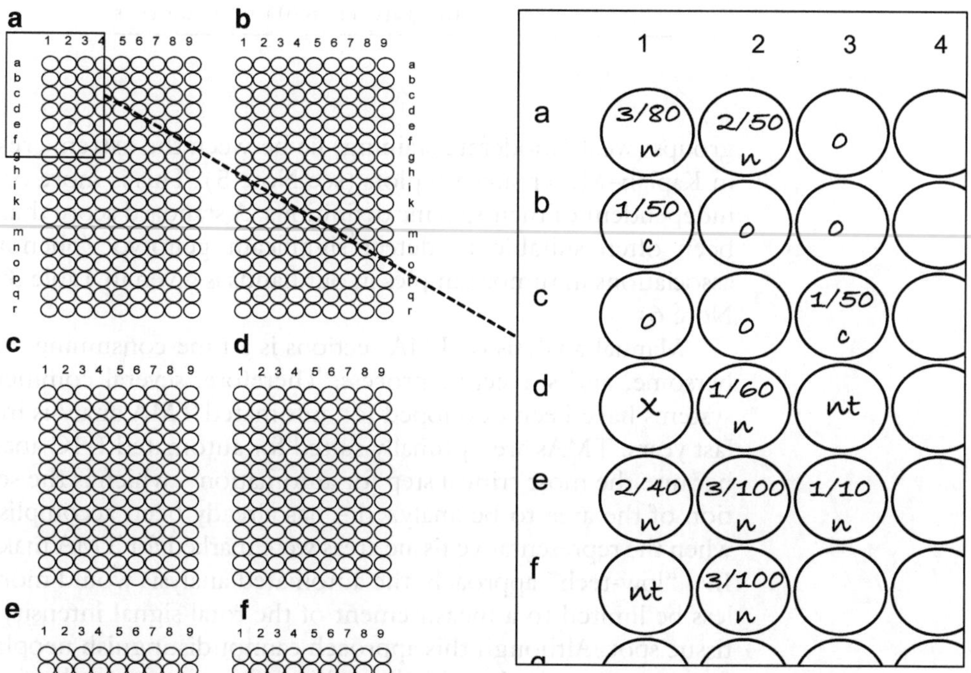

coordinate	IHC intensity	IHC %	IHC pattern	IHC rem	Tissue type	pT	grade	pN
A1a	3	80	nuclear		breast cancer	1	2	0
A1b	1	50	cytoplasmic		breast cancer	2	2	0
A1c	0				breast cancer	2	3	0
A1d				no tissue	breast cancer	3	3	1
A1e	2	40	nuclear		breast cancer	1	1	2
A1f				no tumor	breast cancer	2	2	0

Fig. 4. TMA analysis. Cartoons representing the localization of each tissue spot are helpful to take notes during microscopic analysis. The raw data, including staining intensity (0–3), fraction of stained tumor cells (%), staining patterns (*n* nuclear, *c* cytoplasmic), and remarks (for instance, lack of tissue spots (x), or lack of tumor cells in the spot (*nt* no tumor)) are later transferred into excel spreadsheets.

Table 3
Example of IHC scoring in cancer samples based on the staining intensity (0 = no staining, 3+ = strongest intensity) and the fraction of stained tumor cells

Score	Definition
Negative	Intensity 0
Weak positivity	Intensity 1 in ≤70% of tumor cells or Intensity 2 in ≤30% of tumor cells
Moderate positivity	Intensity 1 in >70% of tumor cells or Intensity 2 in >30% but ≤70% of tumor cells or Intensity 3 in ≤30% of tumor cells
Strong positivity	Intensity 2 in >70% of tumor cells Intensity 3 in >30% of tumor cells

groups (weak, moderate staining) lie between the extreme results in Kaplan–Meyer survival plots (see Note 5). This is more or less independent of their specific definitions. A standard score that has been often suitable to detect significant genotype/phenotype associations in tumor samples in our hands is given in Table 3 (see Note 6).

Manual analysis of TMA sections is a time-consuming, cumbersome, and subjective process. Therefore, several commercial systems have been developed for automated TMA analysis in the last years. TMAs are optimally suited for automated IHC analysis because the most critical step for automation – which is the selection of the area to be analyzed – has already been accomplished when the representative tissue areas were marked for TMA making. In a "low-tech" approach, the automated analysis would more or less be limited to a measurement of the total signal intensity per tissue spot. Although this approach cannot distinguish neoplastic from non-neoplastic epithelial cells or from stroma cells, significant associations with outcome information are usually detected provided an "easy" protein is analyzed. In this case, the high number of samples typically included in TMA analysis compensates for a certain fraction of misinterpreted tissues. Automated analysis with "low-tech" systems may fail, however, if proteins with complex staining patterns are analyzed. For example, the adhesion protein β-catenin shuttles between the cell membrane, cytoplasm, and the nucleus depending on its actual role in cell biology. A low-tech system, which can only measure the intensity (that remains largely unchanged) but cannot identify the expression pattern, will certainly fail to detect this important association. Approaches that can be utilized for such more sophisticated analyses could for example include multicolor fluorescent detection systems for IHC or intelligent pattern recognition software solutions. Such systems come close to the performance of a skilled pathologist (3).

4. Notes

1. If suitable control tissues are not exactly known, it may be advisable to construct a small test TMA containing 50–100 different tissue samples. Such a TMA provides a high likelihood to quickly identify at least one positive tissue sample. Optimally, several positive tissues are identified with varying levels of staining. Consecutive sections of the control tissue (or the test TMA) can be used for protocol optimization.
2. Dilute the secondary antibody to 5 μg/ml in antibody diluent as a starting point.
3. Tissue sections or TMA sections mounted on glass slides are prone to aging. This process may start after a few weeks (4) and results in decreased staining intensity. This is true even when paraffin is not removed from the slides. For best IHC quality, use only freshly cut sections and avoid long-term storage of mounted sections.
4. Manual TMA analysis can be markedly accelerated if a second person enters the results into the TMA ideogram. Up to 1,000 tissue spots per hours can be analyzed this way.
5. Survival plots according to Kaplan–Meier require clinical follow-up data, that is, the time interval between surgery and the event (for instance, tumor recurrence, progression, or patient death), and a censor stating if the event has occurred (censor = 1) or not (censor = 0).
6. Other scores may yield better results. Perform Kaplan–Meier survival analysis to compare the suitability of different scores to define patient groups with different prognosis.

References

1. Shi, S.-R., Cote, R. J. and Taylor, C. R. (2001) Antigen retrieval techniques: current perspectives. J Histochem Cytochem; 8:931–8.
2. Went, P. T., Dirnhofer, S., Bundi, M., Mirlacher, M., Schraml, P., Mangialaio, S., Dimitrijevic, S., Kononen, J., Lugli, A., Simon, R. and Sauter, G. (2004) Prevalence of KIT expression in human tumors. J Clin Oncol; 22:4514–22.
3. Giltnane, J. M., Murren, J. R., Rimm, D. L. and King, B. L. (2006) AQUA and FISH analysis of HER-2/neu expression and amplifica-

tion in a small cell lung carcinoma tissue microarray. Histopathology; 2:161–9.

4. Mirlacher, M., Kasper, M., Storz, M., Knecht, Y., Durmuller, U., Simon, R., Mihatsch, M. J. and Sauter, G. (2004) Influence of slide aging on results of translational research studies using immunohistochemistry. Mod Pathol; 11:1414–20.

5. Tapia, C., Glatz, K., Novotny, H., Lugli, A., Horcic, M., Seemayer, C. A., Tornillo, L., Terracciano, L., Spichtin, H., Mirlacher, M., Simon, R. and Sauter, G. (2007) Close association between HER-2 amplification and overexpression in human tumors of non-breast origin. Mod Pathol; 2:192–8.

Chapter 13

DNA Copy Number Analysis on Tissue Microarrays

Anne Kallioniemi

Abstract

Detection of DNA sequence copy number changes is essential in both clinical practice and basic research, especially in cancer research. The combination of fluorescence in situ hybridization (FISH) and tissue microarray (TMA) technology provides high-throughput means for the evaluation of genetic aberrations in a large number of tissue samples. FISH on TMA is technically demanding and several protocols that include a variety of tissue pretreatment steps have been developed to improve the success of this methodology. Despite of the technical difficulties, FISH analysis on TMA has been successfully used not only to uncover genetic alterations in various malignancies but to also rapidly establish the clinical significance of such changes.

Key words: Tissue microarray, High-throughput screening, Copy number analysis, Fluorescence in situ hybridization, Molecular cytogenetics, Molecular pathology, Genetic aberrations

1. Introduction

Fluorescence in situ hybridization (FISH) has been successfully utilized for many years to study DNA sequence copy number changes, especially in various malignancies. These studies have revealed a wealth of new information on the genetic background of these diseases and have highlighted the importance of specific genetic abnormalities in cancer pathogenesis (1, 2). In some tumor types, FISH technology is routinely used to detect copy number changes affecting specific genes with clinical significance. The most well-known example is the *ERBB2* oncogene which is amplified in a substantial portion of primary breast tumors (3). *ERBB2* amplification is associated with poor patient outcome and either resistance or sensitivity to specific therapies commonly used to treat breast cancer (3, 4). In addition, FISH analysis is frequently used to identify patients that are more likely to benefit from targeted therapy against the *ERBB2* oncogene (3). However,

Ronald Simon (ed.), *Tissue Microarrays: Methods and Protocols*, Methods in Molecular Biology, vol. 664,
DOI 10.1007/978-1-60761-806-5_13, © Springer Science+Business Media, LLC 2010

in majority of cases, FISH remains a research tool that allows direct and accurate measurement of DNA copy number changes in a variety of tissue and cell types.

The advent of the tissue microarray (TMA) technology provided a platform for high throughput copy number analysis in large series of tissue samples (5). FISH analysis on TMAs is especially suitable for rapid determination of the frequency and clinical significance of specific genetic aberrations in large tumor materials (6–8). In the same manner, multitissue arrays containing specimens from a number of different tissues or tumor types can be applied for efficient evaluation of the tissue spectrum of specific genetic changes (9–11). Although FISH analysis on paraffin-embedded tissue specimens is in principle straightforward, the technique is methodologically challenging and its application to TMAs brings along additional technical difficulties. The main problem is that tissue fixation, such as formalin fixation, leads to the formation of cross-linking methylene bridges that are known to adversely influence probe penetration and thereby hybridization efficiency. In the TMA setting, the type and length of tissue fixation as well as the age of the tissues can vary greatly from one specimen to another thus exacerbating these problems. Several protocols aiming to enhance the performance of FISH on TMAs have been published (12–14) and they typically include a variety of pretreatment steps that improve probe penetration. The protocol currently used in our laboratory is detailed below. However, it has to be noted that different tissues will indeed require somewhat different pretreatments and therefore optimization of these steps of the protocol is essential for obtaining successful results. Finally, scoring of hybridization signals from hundreds of tissue specimens is still mainly performed manually and thus remains a tedious and time-consuming task. Automated scoring systems have been developed (14) and will hopefully in the future further increase the power of this technology.

2. Materials

2.1. Probe Labeling

1. BAC DNA (can be isolated using commercial BAC DNA isolation kits).
2. Distilled water.
3. 10× dNTPs: 2 mM dATP, 2 mM dCTP, 2 mM dGTP, 1 mM dTTP in H_2O.
4. 0.5 mM SpectrumOrange-dUTP (Vysis, Des Plaines, IL).
5. 2.5× Random Primers Solution and Klenow Fragment (included in the BioPrime DNA Labeling System, Invitrogen, Carlsbad, CA).

6. Bio-Spin P6 Chromatography Columns (Bio-Rad, Hercules, CA).
7. 3 M Sodium acetate (pH 5.2).
8. 100% Ethanol.
9. TE buffer: 10 mM Tris–HCl, 1 mM EDTA.

2.2. Slide Preparation and Pretreatment

1. Tech Mate microscope slides (DAKO Cytomation, Glostrup, Denmark).
2. Hexan (Scharlaw, Barcelona, Spain).
3. 70, 85, and 100% Ethanol.
4. Deproteination solution: 0.2N HCl.
5. 1× Phosphate Buffered Saline (PBS).
6. Sodium borohydride: 0.3% NaBH4 in 1× PBS.
7. Vysis pretreatment solution (Vysis, Des Plaines, IL).
8. Proteinase treatment solution (prepare solution just prior to use): for each slide to be treated add 1 μl Vysis proteinase (2,500–3,000 U/mg; Vysis, Des Plaines, IL) to 500 μl Vysis proteinase buffer (Vysis, Des Plaines, IL) and preheat to 37°C.
9. Denaturation buffer: 70% formamide, 2× SSC (pH 7.0); store at 4°C.

2.3. Fluorescence In Situ Hybridization

1. Master mix: dissolve 1 g dextran sulfate into 5 ml formamide and 1 ml 20× SSC, adjust pH to 7, and bring total volume to 7 ml with distilled water; store at −20°C.
2. Human Cot-1 DNA (Invitrogen, Carlsbad, CA).
3. Coverslips.
4. Rubber cement.
5. Wash solution I: 0.4× SSC, 0.3% Nonidet P-40.
6. Wash solution II: 2× SSC, 0.1% Nonidet P-40.
7. Counterstain: 0.3 μM DAPI (4′,6-diamimidino-2-phenylindole, Invitrogen, Carlsbad, CA) in Vectashield antifade solution (Vector Laboratories Inc., Burlingame, CA), store in dark at 4°C.

3. Methods

3.1. Probe Labeling

Ready-to-use fluorescently labeled probes are commercially available for certain genes or loci but in most of the cases, custom-made probes are needed. Similar to regular FISH, DNA probes can be generated, for example, from cosmids, P1 artificial chromosomes (PACs), and bacterial artificial chromosomes (BACs). However, in practice large insert size probes, such as PACs and BACs or clone contigs, provide better hybridization signals and are thus preferred.

1. For each labeling reaction, dissolve 200 ng PAC/BAC DNA in 18 μl distilled water in a dark microcentrifuge tube. Denature by heating for 5 min using a boiling water bath or PCR machine and transfer immediately to ice.
2. Add 5 μl 10× dNTPs, 5 μl 0.5 mM SpectrumOrange-dUTP (see Note 1), 20 μl 2.5× Random Primers Solution, and 1 μl Klenow Fragment. Mix gently and spin briefly on a table-top minifuge. Incubate at 37°C for at least 2 h.
3. Use Bio-Spin Chromatography Columns to purify the labeled probes. Follow the manufacturer's instructions and collect the purified probe into microcentrifuge tubes.
4. Precipitate the labeled DNA by adding 5 μl 3 M sodium acetate and 150 μl cold (−20°C) 100% ethanol. Mix gently by inverting the tube and incubate at −20°C for at least an hour (see Note 2). Centrifuge at 15,000 × *g* for 30 min. Carefully remove the supernatant, air dry the pellet (see Note 3), and dissolve into 25 μl TE buffer.
5. Store the labeled probe at −20°C and protect from light.

3.2. Slide Preparation and Pretreatment

1. Cut the required number of 5 μm thick sections from the TMA block using a microtome and carefully place on microscope slides. Place slides on a tray and incubate overnight at 56°C to ensure that the tissues attach to the glass slide (see Note 4).
2. Paraffin has to be removed from the TMA slides prior to hybridization. It is essential that this step will be carried out in a fume hood using appropriate protective laboratory wear. Place 50 ml hexan solution into three coplin jars and 100% ethanol into two coplin jars (see Note 5). Incubate slides in each hexan solution for 10 min and then in the two ethanol solutions for 5 min each. Place slides on a 37° warm plate until they are dry.

   A set of slide pretreatment steps are performed to ensure good probe penetration thereby increasing the hybridization efficiency (see Note 6). The following steps are carried out at room temperature in coplin jars containing 50 ml of the indicated solution unless otherwise noted.
3. Incubate slides in 0.2N HCl deproteination solution for 20 min and then rinse in distilled H_2O for 3 min. Tap slides gently against a paper towel to remove excess liquid.
4. Incubate slides in sodium borohydride solution for 30 min and then rinse in two changes of 1× PBS for 5 min each. Remove excess liquid as above.

5. Place slides into Vysis pretreatment solution preheated to 80°C and incubate for 40 min (see Note 7) and rinse in distilled H_2O for 1 min. Remove excess liquid but do not allow to dry completely.
6. Place slides into a prewarmed (37°C) humidified chamber (see Note 8), add 500 μl proteinase treatment solution on top of each slide, and incubate at 37°C for 20 min. Rinse slides with 1× PBS for 5 min. Remove excess liquid again as above.
7. Prewarm denaturation buffer to 69°C in a water bath and denature slides for 3 min (see Note 7). Transfer slides through a cold (−20°C) ethanol series (70, 85, 100%) for 2 min in each and dry on a warm plate.

3.3. Fluorescence In Situ Hybridization

1. For each slide, prepare hybridization mixture by combining 21 μl Master Mix, 3 μl Cot-1 DNA, and 3–6 μl labeled probes in a dark microcentrifuge tube (see Note 9). Adjust the total volume to 30 μl with distilled H_2O.
2. Incubate the hybridization mixture at 70°C water bath for 5 min to denature the probe DNAs and then transfer the tubes to ice.
3. Pipet the hybridization mixture on the TMA slides, cover with coverslip, and seal the edges with rubber cement. Place the slide into a 37°C humid chamber (see Note 8) and incubate over 1–3 nights at 37°C (see Note 10).
4. The hybridization mixture and unbound probe is removed using a set of washing steps. Place 50 ml of Wash solutions I and II as well as distilled water into separate coplin jars. Preheat the Wash solution I to 72°C (see Note 7); the other two coplin jars are kept at room temperature.
5. Gently remove rubber cement and coverslips from the slides (see Note 11) and place the slides into Wash solution I for 2 min. Transfer slides to Wash solution II for 1 min and then rinse briefly in distilled water to remove any Wash solutions. Air dry the slides in a dark place.
6. Mount the slides and stain the nuclei using DAPI (see Note 12) in an antifade solution. Pipet 10–30 μl counterstain solution (depending on the size of the array) on the slides and cover with coverslip. Hybridization results are examined with a fluorescence microscope equipped with appropriate filters. Slides should be stored in dark at 4°C.

3.4. Analysis

As mentioned in the introduction, hybridization signals are typically scored manually but automated scoring systems are becoming available. A high-quality fluorescence microscope equipped with appropriate filters to visualize different fluorescent signals is used for the evaluation. It is worthwhile to have a map showing the layout of the TMA to assist in the localization of individual specimens and for easy recording of the results. Hybridization signals are counted from a sufficient number of non-overlapping and morphologically intact nuclei. The exact number of nuclei to be counted may vary depending on the application, but one should aim to quantitate 50–100 nuclei for each tissue specimen. It should also be kept in mind that due to the thickness of the tissue sections, the fluorescent signals may be located in different focal planes.

4. Notes

1. Several different fluorescent labels are currently available and different probes can be labeled used different fluorochromes.
2. The precipitation step can be extended to several hours or can be performed overnight.
3. The pellet is not always visible. In case SpectrumOrange-dUTP was used for labeling, the pellet may appear reddish in color.
4. Adhesive-coated tape system can also be used to aid sectioning of the TMA blocks. In this case, the overnight incubation step at 56° should be omitted.
5. Alternatively, xylene solution can also be used for paraffin removal.
6. Different tissues have different hybridization characteristics and the hybridization efficiency is also highly dependent on type and length of original tissue fixation prior to paraffin embedding and the age of the tissues, that is, how long they have been stored in paraffin. Thus, slide pretreatment steps have to be optimized for each set of tissues by varying the concentrations, incubation times, and temperatures. It has to be also kept in mind that all tissues on a given TMA do not have identical hybridization characteristics, and therefore it is often necessary to apply, for example, two different conditions to increase the number of samples that can be successfully evaluated.
7. Warm the solution using a water bath set at a couple of degrees higher than the desired temperature of the solution. Check the temperature of the solution from inside of the coplin jar. Do not place more than two slides into the coplin jar at a time to avoid a drop in the temperature.
8. A plastic box with a lid can be used for this purpose. It is useful to have a metallic block inside the box to provide an even surface for the slides and also to keep the temperature stable.

Place wet tissue paper inside the box to provide moisture. Preheat the box to 37°C before use.

9. The amount of hybridization mixture needed depends on the size of the TMA. Typically 30 μl is sufficient to cover a TMA with an area of 6 cm^2. The amount of labeled probe to be used for each reaction also varies depending on the type of the probe. Increasing the amount of probe will usually improve the intensity of hybridization signals but too much probe will also increase the background. The optimal amount of probe thus needs to be determined experimentally. Combinations of two or three differentially labeled probes can also be used.
10. The hybridization time can be varied and depends on probe quality. In general, a longer hybridization time provides better hybridization signals. However, too long hybridization time can also lead to drying of the slides and thereby to increased background. Commercially available probes typically provide sufficient hybridization signals after an overnight hybridization.
11. If the cover slip does not come off easily, dip the slide briefly into distilled water.
12. Different counterstains can be used. The choice of counterstain depends on the fluorochromes used for probe labeling.

Acknowledgments

The author would like to thank Ms. Eeva Laurila and Ms. Kati Rouhento for their assistance in preparing this article.

References

1. Tibiletti, M. G. (2007) Interphase FISH as a new tool in tumor pathology. *Cytogenet. Genome Res.* **118**, 229–236.
2. Dave, B. J. and Sanger, W. G. (2007) Role of cytogenetics and molecular cytogenetics in the diagnosis of genetic imbalances. *Semin. Pediatr. Neurol.* **14**, 2–6.
3. Ross, J. S., Fletcher, J. A., Linette, G. P., Stec, J., Clark, E., Ayers, M., Symmans, W. F., Pusztai, L., and Bloom, K. J. (2003) The Her-2/neu gene and protein in breast cancer: biomarker and target of therapy. *Oncologist* **8**, 307–325.
4. Nunes, R. A. and Harris, L. N. (2002) The HER2 extracellular domain as a prognostic and predictive factor in breast cancer. *Clin. Breast Cancer* **3**, 125–135.
5. Kononen, J., Bubendorf, L., Kallioniemi, A., Bärlund, M., Schraml, P., Leighton, S., Torhorst, J., Mihatsch, M. J., Sauter, G., and Kallioniemi, O. P. (1998) Tissue microarrays for high-throughput molecular profiling of tumor specimens. *Nat. Med.* **4**, 844–847.
6. Bärlund, M., Forozan, F., Kononen, J., Bubendorf, L., Chen, Y., Bittner, M. L., Torhorst, J., Haas, P., Bucher, C., Sauter, G., Kallioniemi, O. P., and Kallioniemi, A. (2000) Detecting activation of ribosomal protein S6 kinase by complementary DNA and tissue

microarray analysis. *J. Natl. Cancer Inst.* **92**, 1252–1259.

7. Al-Kuraya, K., Schraml, P., Torhorst, J., Tapia, C., Zaharieva, B., Novotny, H., Spichtin, H., Maurer, R., Mirlacher, M., Köchli, O., Zuber, M., Dieterich, H., Mross, F., Wilber, K., Simon, R., and Sauter, G. (2004) Prognostic relevance of gene amplifications and coamplifications in breast cancer. *Cancer Res.* **64**, 8534–8540.
8. Simon, R. and Sauter, G. (2003) Tissue microarray (TMA) applications: implications for molecular medicine. *Expert. Rev. Mol. Med.* **5**, 1–12.
9. Schraml, P., Kononen, J., Bubendorf, L., Moch, H., Bissig, H., Nocito, A., Mihatsch, M. J., Kallioniemi, O. P., and Sauter, G. (1999) Tissue microarrays for gene amplification surveys in many different tumor types. *Clin. Cancer Res.* **5**, 1966–1975.
10. Andersen, C. L., Monni, O., Wagner, U., Kononen, J., Bärlund, M., Bucher, C., Haas, P., Nocito, A., Bissig, H., Sauter, G., and Kallioniemi, A. (2002) High-throughput copy number analysis of 17q23 in 4788 tissue specimens by fluorescence in situ hybridization to tissue microarrays. *Am. J. Pathol.* **161**, 73–79.
11. Hughes-Davies, L., Huntsman, D., Ruas, M., Fuks, F., Bye, J., Chin, S. F., Milner, J., Brown, L. A., Hsu, F., Gilks, B., Nielsen, T., Schulzer, M., Chia, S., Ragaz, J., Cahn, A., Linger, L., Ozdag, H., Cattaneo, E., Jordanova, E. S., Schuuring, E., Yu, D. S., Venkitaraman, A., Ponder, B., Doherty, A., Aparicio, S., Bentley, D., Theillet, C., Ponting, C. P., Caldas, C., and Kouzarides, T. (2003) EMSY links the BRCA2 pathway to sporadic breast and ovarian cancer. *Cell* **115**, 523–535.
12. Andersen, C. L., Hostetter, G., Sauter, G., and Kallioniemi, A. (2001) Improved procedure for fluorescence in situ hybridization on tissue microarrays. *Cytometry.* **45**, 83–86.
13. Chin, S. F., Daigo, Y., Huang, H. E., Iyer, N. G., Callagy, G., Kranjac, T., Gonzalez, M., Sangan, T., Earl, H., and Caldas, C. (2003) A simple and reliable pretreatment protocol facilitates fluorescent in situ hybridisation on tissue microarrays of paraffin wax embedded tumour samples. *Mol. Pathol.* **56**, 275–279.
14. Brown, L. A. and Huntsman, D. (2007) Fluorescent in situ hybridization on tissue microarrays: challenges and solutions. *J. Mol. Histol.* **38**, 151–157.

Chapter 14

RNA Expression Analysis on Formalin-Fixed Paraffin-Embedded Tissues in TMA Format by RNA In Situ Hybridization

Jürgen Veeck and Edgar Dahl

Abstract

The technique of RNA in situ hybridization, i.e., the detection of specific messenger RNA sequences within structurally intact cells or tissues is not widely used in biomedical research, because it can be cumbersome and technically challenging. However, it has a major advantage that warrants and sometimes even requires its application and the associated efforts. RNA in situ hybridization enables a detailed analysis of gene expression in the absence of a suitable antibody to the molecule encoded by the gene of interest. Within the wealth of RNA analysis technologies available nowadays, RNA in situ hybridization still is the only methodology that allows a precise localization of gene expression at a cellular level. This is particularly important if, e.g., new molecular markers or potential drug target molecules have to be analyzed in large cohorts of human tissues. In cancer research, it may be necessary to show that a newly characterized molecule is indeed expressed by the tumor cells themselves, rather than by any surrounding tissue. A protocol is presented here that has been routinely and successfully used on FFPE tissues assembled on a tissue micro array (TMA).

Key words: RNA in situ hybridization, messenger RNA (mRNA), Cellular localization, Target validation, Breast cancer, Tumor marker, Drug target

1. Introduction

RNA in situ hybridization (ISH) allows the precise cellular localization of gene expression in whole tissue sections. Using a single-stranded gene-specific probe, messenger RNA (mRNA) transcripts of interest are detected and visualized directly on a histological tissue section (in situ). The technique of ISH had been principally developed in the late 1960s (1–3), and was refined for mRNA ISH during the 1980s (4). Since then, this method has been continuously optimized by various laboratories (5–9) for different labeling and detection methods. However, the technique is most sensitive and specific when RNA

Ronald Simon (ed.), *Tissue Microarrays: Methods and Protocols*, Methods in Molecular Biology, vol. 664,
DOI 10.1007/978-1-60761-806-5_14, © Springer Science+Business Media, LLC 2010

probes are being used (10, 11). This has two reasons: RNA/RNA hybrids are thermodynamically much more stable than DNA/RNA hybrids and thus, hybridization can be performed at stringent conditions (higher temperature or increased content of formamide in the hybridization buffer). Secondly, RNA/RNA hybrids are resistant to degradation by RNAse A, advantaging the use of this enzyme in order to degrade any unbound (but labeled) probe RNA, which considerably increases the specificity of the detection method.

The suitability of the technique is also considerably influenced by the detection method itself. Radioactive labels, like ^{3}H and ^{35}S, have been very commonly used in the past, and several researchers stay with these protocols due to their yet unmatched sensitivity. Despite of this, in recent years nonradioactive labeling techniques have gained much attention because detection can be both rapid and sensitive, provided an optimized protocol is being applied. The here presented protocol combines high sensitivity and rapidity by using digoxigenin (DIG)-labeled RNA that is detected by an anti-DIG antibody coupled to alkaline phosphatase (AP) (12, 13). Hybridization signals are visualized by permanent dye precipitates using BM-purple.

This protocol can be divided into seven parts. First, a plasmid-based transcription vector containing a cDNA of the gene of interest is linearized and purified. Linearization is done in two different settings in order to generate templates for both antisense (positive probe) and sense (negative control probe) RNA. Next, in vitro transcription is performed by using phage RNA polymerases on these template DNAs. The third step involves the pretreatment of tissue sections and the hybridization of the gene-specific probe. This protocol includes a subtle proteinase-K digestion during pretreatment in order to improve probe penetration into the tissue during hybridization and elutriation of unbound probe during subsequent washing procedures. Step four is posthybridization washing including an RNAse A digestion step to degrade any unbound RNA probe. Step five is the antibody incubation. Step six is the postantibody washing, a seemingly rather unimposing part of the protocol. Continuous washing, however, is of great importance since unbound antibody has to be removed as complete as possible in order to obtain an efficient signal-to-noise ratio. Step seven finally is the detection of the hybridization signals by a dye development procedure.

2. Materials

2.1. Linearization of Plasmid DNA Containing the Gene of Interest

1. IMAGE Consortium cDNA clones are distributed, e.g., by the American Type Culture Collection (Rockville, MD) or ImaGenes GmbH (Berlin, Germany).
2. Adequate restriction enzymes and 10× restriction buffer.

3. RNAse-free $_{dd}H_2O$: double distilled water autoclaved for 1 h at 121°C.
4. DEPC $_{dd}H_2O$: 0.1% (v/v) diethylpyrocarbonate (DEPC) in $_{dd}H_2O$, stir overnight at 37°C, then autoclave for 1 h at 121°C.
5. Tris–HCl-saturated phenol (pH 8.0) (Roti®-Phenol; Carl Roth, Karlsruhe, Germany).
6. Chloroform, molecular biology tested.
7. Sodium acetate: 3 M solution of sodium acetate in RNAse-free $_{dd}H_2O$, pH 7.0.
8. Ethanol: *p.a.* grade for molecular biology.
9. Loading buffer (6× stock): 30% (v/v) glycerol, 6 mM diaminoethanetetraacetic acid, 0.25% (w/v) bromphenol blue; 0.25% (w/v) xylene cyanol. Dissolve components in $_{dd}H_2O$, spin solution at 12,000×*g* for 2 min, store supernatant at 4°C.
10. Tris–boric acid–EDTA buffer (TBE, 10× stock): 890 mM Tris base, 890 mM boric acid, 20 mM EDTA; autoclave for 1 h at 121°C.
11. Agarose: 0.8% agarose (w/v) in 1× TBE buffer, boil and stir constantly, store solution at 60°C to keep solid, if not immediately used.
12. Size marker: SmartLadder 0.2–10 kb.

2.2. In Vitro Transcription

1. DIG RNA Labeling Mix (10× stock, Roche Diagnostics, Mannheim, Germany).
2. Trancription buffer (10× stock, Roche Diagnostics).
3. Protector RNAse inhibitor (20 U/μl, Roche Diagnostics) and RNA polymerases SP6, T7 and/or T3 (20 U/μl, Roche Diagnostics).
4. Lithiumchloride: 4 M LiCl in RNAse-free $_{dd}H_2O$, store solution at room temperature.
5. Ethanol: *p.a.* Grade for molecular biology.
6. PE buffer (10× stock): 100 mM PIPES (pH 6.8) and 10 mM EDTA (pH 8.0), store solution at 4°C.
7. Heparin: 10% solution (w/v) of Heparin in DEPC $_{dd}H_2O$. Store at 4°C.
8. Hybridization buffer: 25 ml formamide (low conductivity), 7.5 ml NaCl 5M, 5 ml PE 10× stock (see Subheading 2.2, step 6), 0.5 ml tRNA (10 mg/ml), 0.25 ml Heparin (10%, see Subheading 2.2, step 7), 0.5 ml bovine serum albumin (10%), and 2.5 ml of sodium dodecyl sulfate (SDS, 20%). Add to a total volume of 50 ml with DEPC $_{dd}H_2O$. Filter buffer

through a 0.45 μm pore size filter. Store at 4°C for immediate use, keep at −20°C for long term storage.

9. MOPS buffer (10× stock): 200 mM 3-[N-Morpholino] propane sulfonic acid (MOPS), 50 mM sodium acetate, and 10 mM EDTA. Adjust to pH 7.0 with NaOH (10N), autoclave solution for 1 h at 121°C, store in the dark (e.g., by wrapping bottle in aluminum foil).
10. Ethidium bromide: 400 μg/ml ethidium bromide in DEPC $_{dd}H_2O$. Store solution in aliquots at −20°C.
11. RNA-agarose (1%): dissolve 0.7 g agarose (SeaKem) in 62 ml of RNAse-free $_{dd}H_2O$ (see Subheading 2.1, step 3), boil and stir constantly. Cool to 50°C, then add 7 ml of 10× MOPS buffer. Prior to pouring the gel into chambers, add 1.75 ml of formaldehyde to the agarose solution and mix. Let gel dry for 1 h at room temperature. Run gel in 350 ml MOPS (1×) buffer at 5 V/cm for 1.5–2 h.
12. RNA size marker: 0.5–10 kb RNA ladder (1 μg/μl).
13. RNA loading buffer (10× stock): 50% (v/v) glycerol, 1 mM EDTA, 0.4% (w/v) bromphenol blue, and 0.4% (w/v) xylene cyanol in DEPC $_{dd}H_2O$.

2.3. Tissue Pretreatment and Hybridization

1. Xylene *p.a.* grade and ethanol *p.a.* grade for molecular biology.
2. Phosphate-buffered saline (PBS, 10× stock): 80 g NaCl, 2 g KCl, 14.4 g $Na_2HPO_4 \times 2\ H_2O$, and 2 g KH_2PO_4, add to 1 l with H_2O and adjust to pH 7.0–7.4. Dissolve 0.1% (v/v) DEPC (see Subheading 2.1, step 4) in the buffer by stirring overnight at 37°C; then autoclave for 1 h at 121°C.
3. Paraformaldehyde (4%): dissolve 4 g of Paraformaldehyde in 90 ml of H_2O, and then add 350 μl of NaOH (2 N). Heat up to 60°C and stir for 15 min. Then add 10 ml of PBS (10×) and 490 μl of HCl (2 N). Adjust to pH 6.8 with NaOH or HCl.
4. Proteinase-K solution: 10 μg/ml proteinase-K, 20 mM Tris–HCl, and 1 mM EDTA in RNAse-free $_{dd}H_2O$ (see Subheading 2.1, step 3), adjust to pH 7.2.
5. Sodium chloride sodium citrate (SSC, 20× stock): 3 M NaCl and 300 mM sodium citrate, adjust to pH 7.0 with HCl, filter solution through a 0.45 μm pore size filter. Dissolve 0.1% (v/v) DEPC (see Subheading 2.1, step 4) in the solution by stirring overnight at 37°C; then autoclave for 1 h at 121°C.
6. Tris–glycine buffer: dissolve 24.5 g Tris base and 15 g glycine (Merck) in 2 l of H_2O. Adjust to pH 7.0 with HCl. Filter solution through a 0.45 μm pore size filter, then autoclave for 1 h at 121°C.

2.4. Posthybridization Washing

1. SSC 2×: dilute from SSC stock solution (20×) in RNAse-free $_{dd}H_2O$ (see Subheading 2.3, step 4)
2. SSC 5×: dilute from SSC stock solution (20×) in RNAse-free $_{dd}H_2O$.
3. Formamide washing solution: mix 120 ml formamide, 15 ml of SSC (20× stock) and add up to 600 ml with RNAse-free $_{dd}H_2O$. Before use, prewarm washing solution to 37°C for 10–15 min. The remaining solution should be stored in the dark.
4. Sodium chloride–Tris-EDTA (NTE) buffer: 0.5 M NaCl, 10 mM Tris–HCl (pH 7.0), and 0.5 mM EDTA in 1 l of RNAse-free $_{dd}H_2O$. Before use, prewarm NTE buffer to 37°C for 10–15 min.
5. RNAse A solution: 25 mg/ml RNAse A in Tris–HCl (pH 7.4). The concentration of the working solution should be 10 μg/ml.

2.5. Antibody Incubation

1. Maleic acid buffer: dissolve 100 mM maleic acid and 150 mM NaCl in RNAse-free $_{dd}H_2O$. Adjust to pH 7.5 with NaOH (10 N). Filter solution through a 0.45 μm pore size filter, then autoclave for 1 h at 121°C.
2. Blocking reagent (10× stock): 10% (w/v) of blocking reagent (Roche Diagnostics) in maleic acid buffer, shortly boil in a microwave oven. Autoclave for 1 h at 121°C, store 50 ml aliquots at −20°C.
3. Blocking solution (1×): dilute blocking reagent (10× stock) 1:10 in maleic acid buffer (see Subheading 2.5, step 2), and then add 0.1% (v/v) Tween-20.
4. Antibody: anti-Digoxigenin-AP, Fab Fragments (0.75 U/μl, Roche Diagnostics).
5. Antibody solution: dilute anti-DIG antibody 1:2,000 with blocking solution (1×) (see Subheading 2.5, step 3). Preincubate antibody dilution for 1 h at 4°C on a laboratory shaker.

2.6. Postantibody Washing

1. TBS (10× stock): 80 g NaCl, 2 g KCl, and 25 ml of 1 M Tris–HCL (pH 7.4), add up to 1 l with $_{dd}H_2O$. Filter solution through a 0.45 μm pore size filter, then autoclave for 1 h at 121°C.
2. TBST: dilute TBS (10× stock) (see Subheading 2.6, step 1) 1:10 in $_{dd}H_2O$. Add 0.1 % (v/v) of Tween-20. Prepare freshly before use.
3. Sodium chloride–Tris–Magnesium chloride-Tween (NTMT): dissolve 100 mM NaCl, 100 mM Tris–HCl (pH 9.5), 50 mM $MgCl_2$, and 0.1% (v/v) Tween-20 in $_{dd}H_2O$. Prepare freshly before use.

2.7. Detection of Hybridization Signals

1. BM-Purple substrate solution: spin BM-Purple (AP substrate precipitating, Roche Diagnostics) for 7 min at 500×*g* in a benchtop centrifuge. Additionally, filter supernatant through a folded filter. Then add final concentrations of 2 mM Levamisol (Sigma-Aldrich) and 0.1% (v/v) Tween-20. Prepare freshly before use. Levamisol, an inhibitor of endogenous alkaline phosphate activity, can be stored as a 200 mM stock solution (100× stock) at −20°C.
2. Nuclear Fast Red solution ('*Kernechtrot*'): for a volume of 100 ml, dissolve 5 g $Al_2(SO_4)_3$ in $_{dd}H_2O$ under moderate heating (50–60°C), then add 100 mg Nuclear Fast Red powder (Merck), stir and heat up (90°C) until completely dissolved. Cool solution down to room temperature and filter through a folded filter.
3. Kaiser's glycerol gelatine: dissolve 10 g of gelatine in 60 ml $_{dd}H_2O$. Then add 70 ml of glycerol. Add 0.1% (w/w) of phenol, stir for 15–25 min under moderate heating (40–50°C) until solution becomes clear. Filter solution through a folded filter. Before use, pre-warm aliquot to 50°C.

3. Methods

3.1. Linearization of Plasmid DNA Containing the Gene of Interest

1. The gene of interest is usually obtained from a public source (e.g., IMAGE clone consortium or ATCC) supplying partial or full length cDNAs cloned into a vector that contains two phage promoter sequences (usually T3 and T7, but also SP6 and either T3 or T7 is common) flanking the multiple cloning sites (see Fig. 1). T3, T7, and SP6 RNA polymerases are phage-specific enzymes that do not interfere with bacterial promoters, thus allowing highly specific in vitro transcription of the target gene. If these cDNA clones have not been sequence-verified by the supplier, they should be sequenced either in-house or externally in order to verify clone identity. Note that there could be several variant forms (e.g., splice variants) of the gene of interest and one should be aware about the kind of transcript variant being analyzed. Plasmid DNA is prepared according to standard procedures (e.g., "Qiagen Maxiprep protocol") and quantified by photometric measurement.
2. RNA in situ hybridization involves the hybridization of two probes, an antisense probe that can hybridize to the mRNA, and a sense probe (negative control) that cannot hybridize. Thus, two different linearization reactions are required (see Fig. 1).

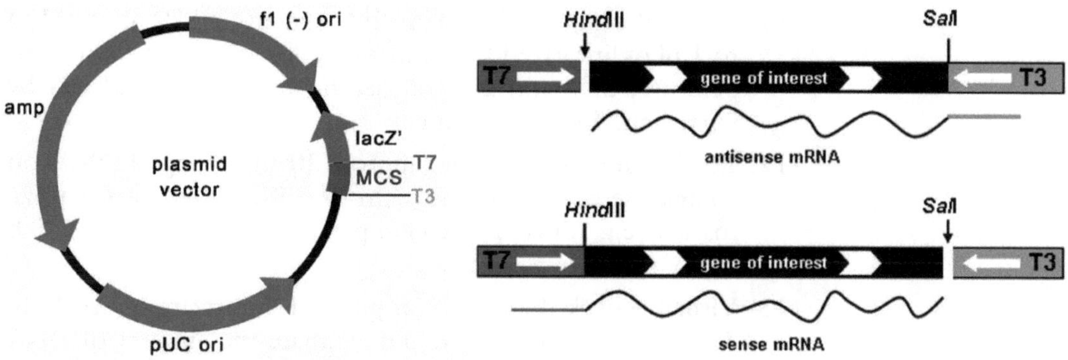

Fig. 1. Principle type of plasmid vectors used for in vitro transcription of RNA. The plasmid vector (e.g., pBluescript SK II) harboring the gene of interest should contain phage promoters such as T3, T7, or SP6 flanking the multiple cloning site (MCS). The plasmid vector is linearized in two separate digest reactions to generate the DNA templates subsequently used for antisense and sense in vitro transcription, respectively. In this example, *Hin*dIII could be used for the T3 dependent in vitro transcription generating antisense mRNA, and *Sal*I could be used in the T7 dependent in vitro transcription to generate sense mRNA. *Arrow* in gene of interest indicates the orientation of the open reading frame.

3. To 20 μg of plasmid DNA, add 10 μl of 10× restriction buffer, add up to 94 μl with RNAse-free $_{dd}H_2O$, and add 6 μl of respective restriction enzyme (10 U/μl). (see Note 1). Incubate for 2 h at the temperature recommended by the enzyme supplier (usually 37°C).
4. Add 100 μl of RNAse-free $_{dd}H_2O$ and 200 μl of Tris–HCl-saturated phenol (pH 8.0) to the reaction tube. Mix by vigorous vortexing (10 s) and spin for 5 min in a benchtop centrifuge (12,000 × *g*, room temperature).
5. Transfer the aqueous supernatant to a new reaction tube; add 200 μl of RNAse-free chloroform. Vortex vigorously (10 s) and spin for 5 min in a benchtop centrifuge (12,000 × *g*, room temperature).
6. Transfer the aqueous supernatant to a new reaction tube and precipitate the DNA with 1/10 volumes of 3 M sodium acetate and 2.5 volumes of ice-cold (−20°C) ethanol abs., vortex for 10 s. Precipitation of DNA is done by incubation for 10 min at −80°C, or alternatively for 2 h up to overnight at −20°C.
7. Spin at least 20 min at 12,000 × *g* in a chilled centrifuge (4°C). Carefully remove the supernatant and discard.
8. Wash the DNA pellet with 400 μl of 70% ethanol (ice-cold) and spin at least 10 min at 12,000 × *g* in a chilled centrifuge (4°C). Carefully remove the supernatant and discard.
9. Dry the DNA pellet by air or in a vacuum (e.g., in speed-vac) and then resuspend the DNA in 40 μl of DEPC $_{dd}H_2O$.
10. Keep samples on ice or store at −20°C, if the in vitro transcription is not being performed on the same day.

11. For verification of the obtained fragments, add 9 µl of $_{dd}H_2O$ to 1 µl of linearized DNA, incubate for 15 min at room temperature, then add 2 µl of loading buffer (6×) and mix by vortexing. Spin down shortly.
12. Load 1 and 2 µl per tube (approx. 20 and 40 ng of DNA) on an agarose gel (0.8% agarose in 1× TBE-buffer). Load a size marker (e.g., SmartLadder) in parallel. Run the gel for 2–3 h at 5 V/cm (see Note 2).
13. Visually check the quality of plasmid linearization (see Note 3), the size of the DNA, and estimate the amount of DNA still present in the preparation.

3.2. In Vitro Transcription

1. To 1 µg of linearized DNA template (see Subheading 3.1) add RNAse-free $_{dd}H_2O$ up to a volume of 14 µl (see Note 4). Add 2 µl of RNA Labeling Mix (contains the label Digoxigenin-UTP), 2 µl of 10× transcription buffer, 0.8 µl of RNAse inhibitor and 1.2 µl of the respective RNA polymerase (T3, T7 or SP6). Mix by pipetting, spin down shortly. Incubate at 37°C for 2 h.
2. Precipitate the in vitro transcribed RNA by adding 2.5 µl of 4 M LiCl and 75 µl of ice-cold (–20°C) 100% ethanol. Incubate for 10 min at –80°C.
3. Centrifuge for 30 min at 12,000×*g* in a chilled benchtop centrifuge (4°C).
4. Remove the supernatant and wash the RNA pellet with 400 µl of 70% ethanol (–20°C). Centrifuge for 10 min at 12,000×*g* in a chilled benchtop centrifuge (4°C).
5. Carefully remove the supernatant and let the RNA pellet dry at room temperature for 5 min.
6. Dissolve the RNA pellet in 51 µl of DEPC $_{dd}H_2O$. Transfer 2.5 µl of the RNA solution into an extra reaction tube on ice for subsequent RNA gel electrophoresis (see Subheading 3.2, step 9).
7. The remaining 48.5 µl are distributed on two reaction tubes (each 24 µl). To each tube, add 456 µl of hybridization buffer (1:20 dilution).
8. RNA is stored until use in hybridization buffer at –20°C. The RNA concentration of this preparation should be approximately 10 ng/µl (see Note 5).
9. To prepare an RNA gel (here 70 ml volume; adjust volumes to your gel sizes), add 62 ml of RNAse-free $_{dd}H_2O$ to 0.7 g of agarose, boil in a microwave oven to dissolve, cool down to 50°C, and then add 7 ml of 10× MOPS buffer.
10. Add 1.75 ml of formaldehyde (37%), mix by turning, and directly pour the gel before the gel solution gets solid. Let the gel dry for 1 h.

11. Put the RNA gel in the electrophoresis chamber and add sufficient 1× MOPS buffer for gel electrophoresis. Ideally, the gel is covered by no more than 2 mm of buffer.
12. To 2.5 μl of in vitro transcribed RNA (and to 4 μl of the RNA-ladder), add the following: 5 μl formamide (low conductivity), 2 μl formaldehyde, 1 μl ethidium bromide, 1 μl 1× MOPS (see Note 6).
13. Incubate for 10 min at 65°C to denature RNA secondary and ternary structures. Put on ice to prevent RNA renaturation. Add 2 μl of loading buffer (6×). Load the complete content on an RNA gel and perform electrophoresis for 2 h at 5 V/cm.

3.3. Tissue Pretreatment and Hybridization

1. Perform all pretreatment steps in glass cuvettes (200 ml) under a flue at room temperature. Any glassware should be sterilized at 200°C for 4–5 h before use.
2. Transfer the Formalin-Fixed Paraffin-Embedded (FFPE) tissue slides to be hybridized through the series of solvents and solutions as indicated in Table 1.
3. Prepare the complete hybridization solution *before* you take out the slides from the Tris–glycine buffer. Hybridization solution with antisense and sense probes (10 ng/μl), respectively, that has been stored at −20°C (see Subheading 3.2, step 8) is warmed to 60°C until the solution gets completely clear. Antisense and sense probes are hybridized at a concentration of 2.5 ng/μl (see Note 7), thus the probes have to be diluted in prewarmed hybridization buffer (60°C) 1:4 before applying this solution.
4. Take slides one by one from the Tris–glycine buffer and process singularly as the slides are not allowed to dry out. Remove excess fluid on slides with a KimWipe paper without touching the tissue parts (see Note 8).
5. Apply hybridization solution to the still wet tissue on the slide. Take care not to hurt the tissue with the pipette tip. Cover the tissue with a clean strip of Parafilm® (Brand, Wertheim, Germany) (cut before to a size of 24×40 mm) (see Note 9).
6. Put slides into a 4-chamber dish, wrap the 4-chamber dish with wet paper and put the wrapped dishes in a large plastic box that can be sealed with a lid (see Note 10).
7. Incubate slides with hybridization buffer overnight at 65°C in a shaking water bath.

3.4. Posthybridization Washing

1. Transfer slides into 5× SSC at room temperature and wait until Parafilm® strips float off by themselves. Fish away the Parafilm® strips with a forceps. Wash for further 20 min. Repeat washing two times with fresh 5× SSC.

Table 1
Dewaxing, Rehydration, and Permeabilization of FFPE Tissues

Solvent and solution	Duration (min)	Frequency
Xylene	10	3×
100% Ethanol	5	2×
95% Ethanol	1	1×
90% Ethanol	1	1×
80% Ethanol	1	1×
70% Ethanol	1	1×
50% Ethanol	1	1×
30% Ethanol	1	1×
PBS (1×)	5	2×
4% PFA	30	1×
PBS (1×)	5	2×
Proteinase-K	10	1×
PBS (1×)	5	1×
4% PFA	30	1×
PBS (1×)	5	2×
SSC (2×)	2	2×
Tris–glycine buffer	≥30	1×

2. Incubate slides for 40 min in formamide washing solution at 60°C in a shaking water bath.
3. Transfer slides into fresh formamide washing solution (preheated to 60°C) and transfer to a 37°C shaking water bath. Cooling down of the washing solution will take 1.5 h (60°C → 37°C)
4. Wash slides in NTE buffer for 15 min at 37°C.
5. Digest unbound RNA in NTE buffer containing RNAse A (10 μg/ml) for 30 min at 37°C (see Note 11).
6. Wash slides in NTE buffer for 15 min at 37°C.
7. Incubate slides in formamide washing solution for 40 min at 60°C.
8. Wash slides twice in 2× SSC for 5 min at room temperature.

3.5. Antibody Incubation

1. Preincubate slides in 1% (w/v) blocking solution at room temperature for at least 1 h before the anti-DIG antibody is added. Preincubation for several hours is fine as well.

2. In parallel, dilute the anti-DIG antibody 1:2,000 in 1% blocking solution and mix on a laboratory shaker at 4°C for at least 1 h. Preincubation for several hours is fine as well (see Note 12).
3. Take slides out of the blocking solution one by one, i.e., process singularly so they will not dry out. Remove excess fluid around the tissue with KimWipe tissues.
4. A small Parafilm® ball (diameter approximately 1–2 mm) is put on the shaded part of the slides. Slides are laid upside-down in the holes of a 4-chamber dish (see Fig. 2).
5. In each chamber, 900 μl of diluted antibody solution is pipetted between slide and the bottom of the chamber.
6. The 4-chamber dishes are wrapped in wet paper tissues, put in a large plastic box with lid and incubated overnight at 4°C. No agitation is necessary during antibody incubation.

3.6. Postantibody Washing

1. All washing steps after the antibody incubation are performed on a laboratory shaker.
2. Slides are removed from the 4-chamber dishes with a forceps and directly transferred to a washing solution.
3. Wash slides four times in 1× TBST (each 10 min).
4. Wash slides three times in 1× TBST (each 1 h).
5. Alternatively to Subheading 3.6, step 4, slides may be washed overnight or over the weekend in 1× TBST at 4°C without agitation and without changing the washing solution (see Note 13).
6. Wash slides three times in 1× NTMT (each 10 min).

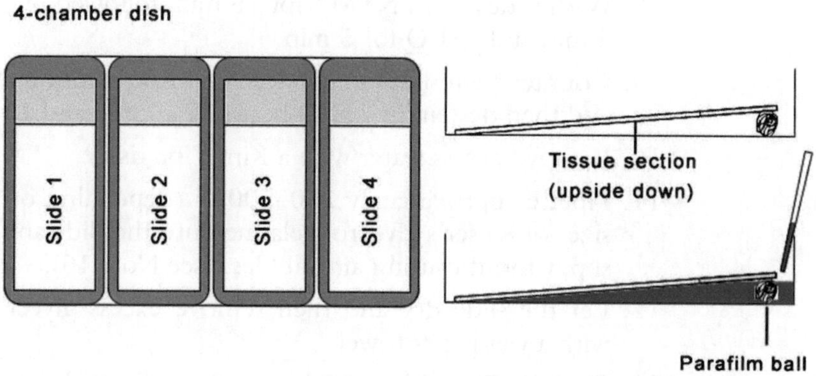

Fig. 2. Scheme of tissue slides accommodation in 4-chamber dishes during antibody incubation and the colorimetric staining reaction. Slides are lying upside down on a small ball (diameter of 1–2 mm) formed from Parafilm® that is chemical inert to the substances applied in this protocol. The design minimizes fluid volumes but prevents drying and damage of the tissue.

3.7. Detection of Hybridization Signals

1. Slides are taken one by one out of the NTMT buffer (processed singularly to prevent drying). Remove excess fluid around the tissue with KimWipe tissues.
2. A small Parafilm® ball (diameter roughly 1–2 mm) is put on the shaded part of the slides. Slides are laid upside-down in the holes of a 4-chamber dish (see Fig. 2).
3. To each chamber, 1 ml of BM-purple solution is pipetted between slide and the bottom of the chamber.
4. The 4-chamber dish are wrapped in wet paper tissues and put in a large plastic box that can be sealed with a lid. Staining should be done in the dark at 4°C, e.g., by putting the box in a refrigerator.
5. On the first day, staining may be also performed at room temperature. Every 2 h staining is checked by two ways: (1) Put a white piece of paper below the staining tray and consider the general signal strength in the sections with the antisense probe compared to the sections with the sense control. Clear differences should become visible with increasing staining time. (2) Look at the staining under an inverse microscope used for cell culture, e.g., Axiovert 25 (Zeiss, Jena, Germany). This will help to distinguish specific staining from unspecific staining that may also be visible in the sense controls (see Fig. 3).
6. Staining is generally done for 1–7 days depending on the strength of the signal (see Note 14). Staining should be stopped when the signal strength in the antisense slides is sufficient and the background in the sense slides is still low (see Note 15).
7. Slides are taken out of the staining solution using a forceps and transferred into NTMT buffer in which slides can be stored for several days.
8. Wash slides 2× in NTMT for 15 min, followed by 1× PBS for 5 min and $_{dd}H_2O$ for 5 min.
9. Counter stain slides in Nuclear Fast Red solution for 10 min, and then destain in $_{dd}H_2O$ between 10 min and 1 h.
10. Remove excess water with a KimWipe tissue.
11. Pipette approximately 250–500 μl (depending on the tissue size) of Kaiser's glycerol gelatine onto the slide and add cover slip without causing air bubbles. (see Note 16).
12. Let the slide dry and then remove excess glycerin-gelatine with a wet paper towel.
13. DONE. The slides are now ready to be analyzed under the microscope. Figure 4 shows an example of RNA in situ hybridization results obtained with this protocol.

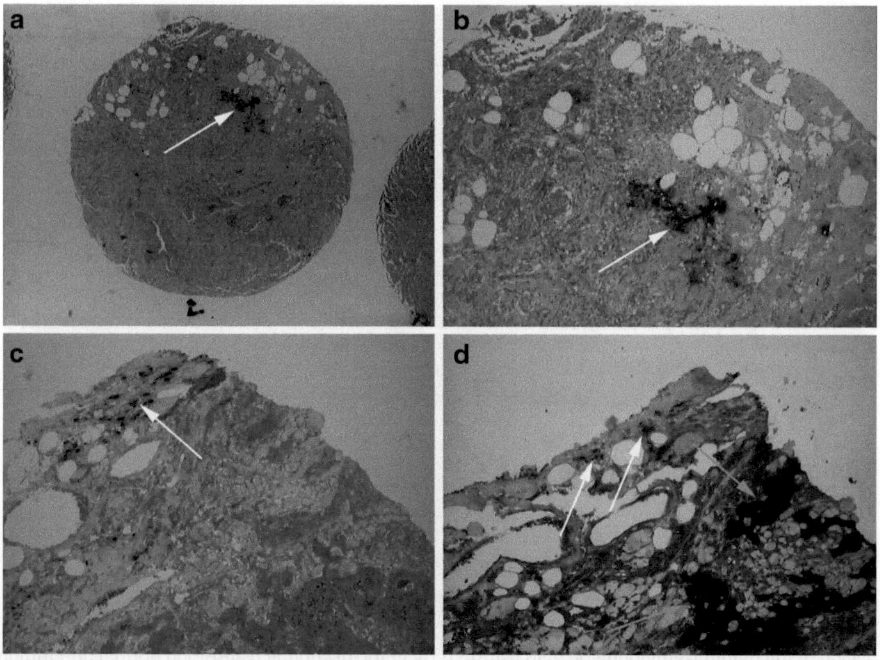

Fig. 3. Examples of unspecific staining that has to be controlled by the sense RNA hybridization. Since no RNA in situ hybridization protocol is without risk for unspecific signals, analysis of the sense control is mandatory. (**a**) Typical example of unspecific signals in a nonoptimal ISH probed with sense RNA: Unspecific staining (*white arrow*) is sometimes found near sites of tissue damage or tissue folding. (**b**) Magnification from sample shown in (**a**). (**c** and **d**): High magnification of unspecific staining at the edges of tissue (*two left arrows*) can be distinguished from specific staining (*two right arrows* in **d**) with sense (**c**) and antisense (**d**) probes. Magnifications: (**a**), ×40; (**b–d**): ×100.

4. Notes

1. For in vitro transcription, the choice of restriction enzyme used for linearization is important: enzymes creating 3′-overhangs, like *Pst*I or *Sac*I, should not be used because these overhangs are thought to interfere with RNA polymerases (14).
2. This is the first step in the protocol which can be "frustrating." After phenol-chloroform extraction, linearized DNA dissolved in water tends to appear as multiple bands though linearization was successful. According to our experience, this artefact can be overcome if the DNA is preincubated in a salt buffer (1× restriction enzyme buffer) for 15 min before gel electrophoresis.
3. If DNA has not been linearized completely, repeat the restriction digest again on the partially digested DNA but run in parallel a fresh sample of plasmid DNA (10 μg). It may be helpful to choose a fresh batch of restriction enzyme.

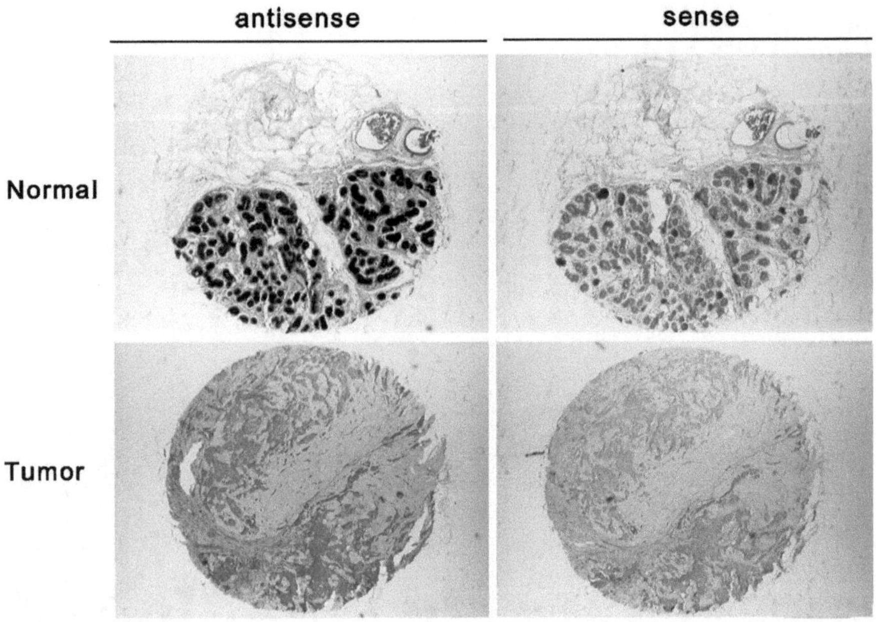

Fig. 4. Expression of the gamma-aminobutyric acid (GABA) A receptor, pi subunit (GABRP) in normal and tumorous breast tissue. FFPE tissue core sections (3 μm) from a TMA were analyzed for GABRP mRNA expression using the presented protocol. Staining with the gene-specific antisense-RNA probe reveals very abundant GABRP mRNA expression in normal breast epithelial cells, while absence of expression is seen in surrounding stromal cells, endothelial cells and connective tissue. The absence of staining in the negative control (normal tissue probed with a GABRP-specific sense RNA) demonstrates the high specificity of the antisense probe for the detection of GABRP mRNA expression. However, in a high percentage of human breast cancer specimens GABRP expression is lost (15). In these breast tumors, GABRP mRNA expression is not detectable with the antisense probe and the background staining is comparable to that of the negative control. Magnifications: ×40.

4. If the concentration of the linearized DNA template is low so that 14 μl do not equal 1 μg, the whole set-up is doubled having the DNA in a maximal volume of 28 μl.
5. In vitro transcription from 1 μg DNA template is thought to generate roughly 10 μg of RNA. Thus, 2.5 μl for the RNA gel electrophoresis should represent 500 ng of RNA that should be easily visible. After a 1:20 dilution in hybridization buffer, the probes have an RNA concentration of approximately 10 ng/μl.
6. The RNA additives formamide (low conductivity), formaldehyde, ethidium bromide, and 1× MOPS can be stored as a "mastermix" for several months at –20°C.
7. Two and a half ng/μl is actually a starting point to get a signal anyway. Specific and sensitive antisense probe should detect their mRNA also very well at a concentration of 1.0 ng/μl or less which is advantageous because background staining will be very low.
8. It is important to have a dry slide area around the wet (biological) tissue; otherwise the hybridization cocktail could

smear and will be diluted too much. KimWipe paper towels are used because they do not get "fuzzy". These papers are definitely not RNAse-free; however, they are regarded as RNAse-free in this protocol. Wear fresh gloves when you do this procedure.

9. Take care that these Parafilm® strips are close to the mentioned size and prepare in advance to the hybridization procedure. If strips are too big, the hybridization solution may leak out due to adhesion bridges to dish material.
10. The sealed lid is important so that wrapped dishes may not dry out, and no condensation water from the shaking water bath may enter.
11. Though RNAse on glassware is thought to be inactivated after several hours of baking we routinely used a separately labeled glass cuvette for the RNAse A digest.
12. This preincubation is thought to dilute out (by binding to blocking solution components) unspecific antibodies potentially present in the anti-DIG antibody preparation.
13. The duration of washing in the 1× TBST buffer is critical for the success of the experiment. Insufficient washing will cause increased background, meaning that the staining of slides has to be stopped prematurely. According to our experience, washing can be done for up to 1 week, however, at 4°C, in order to retain the enzymatic activity of the alkaline phosphatase.
14. The staining solution should be changed every day. However, if the signal that develops is weak, it is sufficient to change the staining solution every second day.
15. Staining intensity under the microscope usually does not look as strong as staining intensity considered by the "white paper method." Therefore, sections may be slightly "overstained." To detect expression in low expressing cells, check that the sense probe is negative in the same region of the tissue.
16. If a mounting medium other than the recommended is used, the applier should be aware that xylene-based mounting media support the formation of crystal structures from the staining precipitate.

Acknowledgments

The authors would like to thank the former and present lab technicians Annette Buß, Beate Petschke, and Sevim Alkaya, who constantly helped to further optimize this RNA in situ hybridization protocol over the years.

References

1. Gall JG, Pardue ML (1969). Formation and detection of RNA-DNA hybrid molecules in cytological preparations. Proc Natl Acad Sci USA;63:378–383.
2. John HA, Birnstiel ML, Jones KW (1969). RNA–DNA hybrids at the cytological level. Nature;223:582–587.
3. Buongiorno-Nardelli M, Amaldi F (1970). Autoradiographic detection of molecular hybrids between rRNA and DNA in tissue sections. Nature;225:946–947.
4. Angerer LM, Angerer RC (1981). Detection of poly A+ RNA in sea urchin eggs and embryos by quantitative in situ hybridization. Nucleic Acids Res;9:2819–2840.
5. Miller MA, Kolb PE, Raskind MA (1993). A method for simultaneous detection of multiple mRNAs using digoxigenin and radioisotopic cRNA probes. J Histochem Cytochem; 41:1741–1750.
6. Pohle T, Shahin M, Gillessen A, Schuppan D, Herbst H, Domschke W (1996). Expression of type I and IV collagen mRNAs in healing gastric ulcers – a comparative analysis using isotopic and non-radioactive in situ hybridization. Histochem Cell Biol;106: 413–418.
7. Renz M, Kurz C (1984). A colorimetric method for DNA hybridization. Nucleic Acids Res;12:3435–3444.
8. Wisden W, Morris B, Hunt S (1991). Molecular Neurobiology, A. Practical Approach (eds. J. Chad and H. Wheal), pp. 205–225. University Press/IRL, Oxford.
9. Dahl E, Winterhager E, Traub O, Willecke K (1995). Expression of gap junction genes, connexin40 and connexin43, during fetal mouse development. Anat Embryol (Berl); 191:267–278.
10. Cox KH, DeLeon DV, Angerer LM, Angerer RC (1984). Detection of mrnas in sea urchin embryos by in situ hybridization using asymmetric RNA probes. Dev Biol;101:485–502.
11. Feinberg AP, Vogelstein B (1983). A technique for radiolabeling DNA restriction endonuclease fragments to high specific activity. Anal Biochem;132:6–13.
12. Tautz D, Pfeifle C (1989). A non-radioactive in situ hybridization method for the localization of specific RNAs in *Drosophila* embryos reveals translational control of the segmentation gene hunchback. Chromosoma;98:81–85.
13. Höltke HJ, Kessler C (1990). Non-radioactive labeling of RNA transcripts in vitro with the hapten digoxigenin (DIG); hybridization and ELISA-based detection. Nucleic Acids Res;18:5843–5851.
14. Gilman M (1993). Current Protocols in Molecular Biology (eds. Ausubel FM, Brent R, Kingston RE, Moore DD, Seidman JG, Smith JA, Struhl K), Vol. 1, pp. 4.7.1–4.7.6, Greene and Wiley-Interscience, New York.
15. Zafrakas M, Chorovicer M, Klaman I, Kristiansen G, Wild PJ, Heindrichs U, Knüchel R, Dahl E (2006). Systematic characterisation of GABRP expression in sporadic breast cancer and normal breast tissue. Int J Cancer;118:1453–1459.

Chapter 15

Automated Analysis of Tissue Microarrays

Marisa Dolled-Filhart, Mark Gustavson, Robert L. Camp, David L. Rimm, John L. Tonkinson, and Jason Christiansen

Abstract

The analysis of protein expression in tissue by immunohistochemistry (IHC) presents three significant challenges. They are (1) the time-consuming nature of pathologist-based scoring of slides; (2) the need for objective quantification and localization of protein expression; and (3) the need for a highly reproducible measurement to limit intra- and inter-observer variability. While there are a variety of commercially available platforms for automated chromagen-based and fluorescence-based image acquisition of tissue microarrays, this chapter is focused on the analysis of fluorescent images by AQUA® analysis (*A*utomated *QU*antitative *A*nalysis) and the solutions offered by such a method for research and diagnostics. AQUA analysis is a method for molecularly defining regions of interest or "compartments" within a tissue section. The methodology can be utilized with tissue microarrays to provide rapid, quantitative, localized, and reproducible protein expression data that can then be used to identify statistically relevant correlations in populations. Ultimately this allows for a multiplexed, objective and standardized quantitative approach for biomarker research and diagnostic assay development for protein expression in tissue.

Key words: Immunohistochemistry, Automated analysis, AQUA, Tissue microarrays, Quantitative analysis, Biomarkers, Immunofluorescence

1. Introduction

1.1. Automated Immunostaining Evaluation of Tissue Microarrays

Formalin-fixed paraffin embedded (FFPE) tissue samples are commonly used for IHC analysis due to stability that allows for long-term tissue archiving (1–3) as well as maintenance of morphologic features of the sample. This is in contrast to other methods such as ELISA, flow cytometry, and western blotting which require disaggregating tumor samples resulting in both the loss of tissue architecture and the potential inclusion of adjacent non-tumor regions. The prevalence of FFPE tumor archives serves as

Ronald Simon (ed.), *Tissue Microarrays: Methods and Protocols*, Methods in Molecular Biology, vol. 664,
DOI 10.1007/978-1-60761-806-5_15, © Springer Science+Business Media, LLC 2010

a rich resource for investigators for the testing and validation of hypotheses through the construction of tissue microarrays for examination of large cohorts in a rapid time frame. In addition, just as histopathological assessment has remained the gold-standard for diagnosis of patient samples, IHC has become a key research tool in that it allows for the examination of protein expression levels and localization while preserving the sample morphology.

Classic IHC analysis as originally implemented in diagnostic pathology and histopathological evaluation relies on the deposition of a chromagen substrate such as 3,3′-diaminobenzidine (DAB) at the site of antibody/antigen binding. The sample is then analyzed by a pathologist using light microscopy in the context of a counterstain (such as H&E) for determination of a binary variable – assessment of whether the protein is "present" or "not present". However, this simple analysis does not account for the true range of protein expression across a cohort of samples (4) nor does it allow for accurate and objective assessments of the degree of chromagen staining within a sample which is directly related to protein expression (5). Fluorescent analysis allows for a broad dynamic range of detection and for a continuous rather than discreet range of expression measurement.

The use of immunofluorescence (IF) for evaluation of protein expression levels has the additional advantages of allowing multiple markers to be measured on the same slide (multiplexing), and for the application of image analysis algorithms for localization and quantification of protein expression over a wider dynamic range than its traditional IHC counterpart (5). IF techniques also permit the co-localization of protein biomarkers. There are several disadvantages to IF-based assessment of protein expression such as potential cross-reactivity of multiple antibodies used to probe a single sample, as well as less than optimal signal-to-noise ratios (low target signal intensities in a background of endogenous tissue autofluorescence). However, the use of antibody optimization/testing, enzymatic amplification of target signals (e.g., using tyramide) and the use of far-red fluorophores can overcome these potential complications. The net advantage of using quantitative IF has been demonstrated on multiple platforms and has been shown to be the measurement of subtle differences in protein expression missed by classical IHC scoring (5–9).

Measurement of protein biomarkers at the site of disease, i.e., within a tumor, may prove to be a useful tool for the development of diagnostic assays for guiding therapeutic decisions. However, the hurdle is very high for the use and application of IHC assays as a tool for making patient treatment decisions (10). Most significantly, the problems lie with making a reproducible and standardized measurement in order to have reliable clinical decisions. Studies have shown that reproducibility is problematic across multiple pathologists and multiple testing sites (inter- and intra- laboratory variation, central vs. non-centralized testing) (11). Sources of variation are introduced due to sample handling (fixation time and methods), sample treatment (type and timing of antigen retrieval methods) as well as the subjective nature of semi-quantitative IHC evaluation.

Part of the variability of IHC analysis can be overcome through the use of tissue microarrays. Because all of the samples are on a single slide for antigen retrieval and antibody binding and are thus treated uniformly, the inherent variation is minimized. To date, traditional IHC results generated by pathologist scoring of chromagen presence include scoring methods such as the basic categorization of chromagen staining to a scale of 0 (no staining), 1+ (weak staining), 2+ (moderate staining), and 3+ (strong intensity), as well as the Allred score and H-score methods that take into account percent positivity of the sample (12, 13). Given the small area of sampling represented by tissue microarray coring, the percent positivity may not be taken into account in tissue microarray scoring. Despite significant inter- and intra-observer variability, this analysis method currently represents the standard of care for pathological evaluation, but these deficiencies set the stage for a computer to objectively determine the intensity of protein expression by IHC analysis.

1.2. Automated Image Acquisition of Tissue Microarrays

The precise row-column grid format utilized in tissue microarray construction is ideal for adaptation to automated image acquisition and analysis. This is evidenced by the plethora of imaging platforms listed in Table 1 for brightfield and/or fluorescence digitization of tissue microarray spots. These platforms generate digital files that can be utilized for image sharing, archiving, and for image analysis.

Many of the systems listed in Table 1 offer image analysis of digitized tissue microarray cores following image acquisition with their platforms based on combinations of morphometric evaluation, feature extraction, pattern recognition and chromagenic signal assessment by density analysis and/or counting based on optical density thresholds. Several also offer digital storage of images in databases and the ability to view and share images through the web.

1.3. Methodology of Automated Quantitative Analysis (AQUA® Analysis): Post Acquisition

AQUA analysis was developed at Yale University for measurement of proteins in situ (14) and has been commercialized by HistoRx, Inc. Rather than morphometric or feature-extraction-based analysis, AQUA analysis is based on molecular co-localization of proteins with cell types and/or subcellular organelles of interest. This methodology has been applied to the study of protein expression in many different tissue types. Some of the applications of AQUA analysis include: predicting patient response and/or outcome by measurement of biomarkers (15), antibody screening

Table 1
Commercially available platforms for automated tissue microarray imaging

Company	Product	Website
Aperio	Scanscope®	http://www.aperio.com
Biogenex	iVision™ and GenoMx™	http://www.biogenex.com
Bioimagene	PATHIAM™	http://www.bioimagene.com
Compucyte corporation	LSC® and iCyte®	http://www.compucyte.com
Dako cytomation	ACIS®	http://www.dako.com
Dmetrix	Dx-40	http://www.dmetrix.net
Genetix	Applied Imaging Ariol®	http://www.genetix.com
Hamamatsu	Nanozoomer	http://www.hamamatsu.com
HistoRx	AQUA®/PM2000™	http://www.historx.com
Molecular devices	Discovery-1™	http://www.moleculardevices.com
Olympus	BLISS HD™	http://olympusamerica.com
TissueGnostics	TissueFAXS	http://www.tissuegnostics.com

and validation (16), validation of transcriptional profiling studies (17), prediction of disease recurrence (18) discrimination between different tumor subtypes (19), and prediction of response to therapy (8, 15).

Although AQUA analysis represents a fundamental advance in the quantification of tissue-based protein biomarkers, the data generated have been shown to correlate with traditional IHC and ELISA data (14, 20). The key differentiating factor is the ability to be quantitative and reproducible on tissue. It is also clear that the localization of biomarkers to different subcellular compartments can be critical in analyzing their potential role as tumor biomarkers, as has been shown for beta-catenin and other biomarkers (14, 18, 21, 22). This type of organelle-specific measurement is not possible using chromagenic-based qualitative techniques.

An overview of the general immunofluorescent staining methodology for use with AQUA analysis is shown in Fig. 1, with its methodology and protocol recently described in more detail (23). The work flow between traditional chromagen-based staining and IF staining utilized for AQUA analysis is very similar, obviating the need for specialized equipment for staining slides. The key difference is the use of multiple molecular species with AQUA analysis to molecularly define cellular and sub-cellular compartments of interest. These reagents are "batched" together in a cocktail.

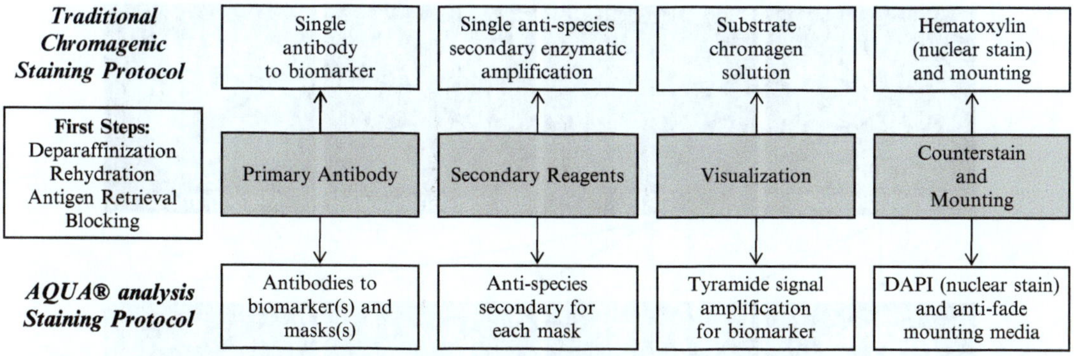

Fig. 1. Comparison between traditional chromagenic IHC staining protocol and IF staining protocol for AQUA® analysis. A diagram of the typical work-flow for IF staining, imaging and use in AQUA analysis (*bottom*) follows the same sequential steps as traditional IHC staining utilizing chromagen-based detection (*top*).

An example of this multiplexing in a breast cancer specimen is shown in Fig. 2. In this experiment, the sample was stained with an anti-pan-cytokeratin antibody (to molecularly identify epithelial cells and non-nuclear regions) and DAPI (to molecularly identify nuclei), as well as with an antibody to the biomarker of interest. Other types of compartments can be analyzed such as stromal regions or other organelles, as well as cell types that can be identified with specific molecular markers (24, 25).

In AQUA analysis, the molecular identification of compartments results in virtual "masks" being created to integrate the area of region of interest. The selection of masks used in a given experiment is determined by the molecular markers (antibodies or chemical entities) utilized that are chosen based on the experimental design and biological hypothesis. This is in contrast to other methodologies that utilize morphometric criteria for defining subcellular compartments and require software to be trained according to physical and mathematical parameters. Utilizing AQUA, the staining intensity of the biomarker being studied (detected by an antibody) is measured and then normalized for the area of compartment of interest as defined by the mask. Normalization of protein expression to the actual area of expression is critical because many tumors contain significant amounts of host-derived desmoplastic stroma. If these areas were included in the analysis, a significantly skewed data set would result. However, AQUA can be utilized to measure biomarkers within such regions of interest as well by the use of appropriate reagents. The value is also normalized for exposure time (the time of sample illumination for the fluorescent measurement to be taken) so that a wide range of protein concentrations can be studied. The result is an AQUA score – a continuous linear variable that provides a quantitative unit of measure. In addition, the use of the Validator™ program with AQUA analysis allows for automated image validation which also reduces the time demands associated with image-by-image inspection. This process is described step by step below.

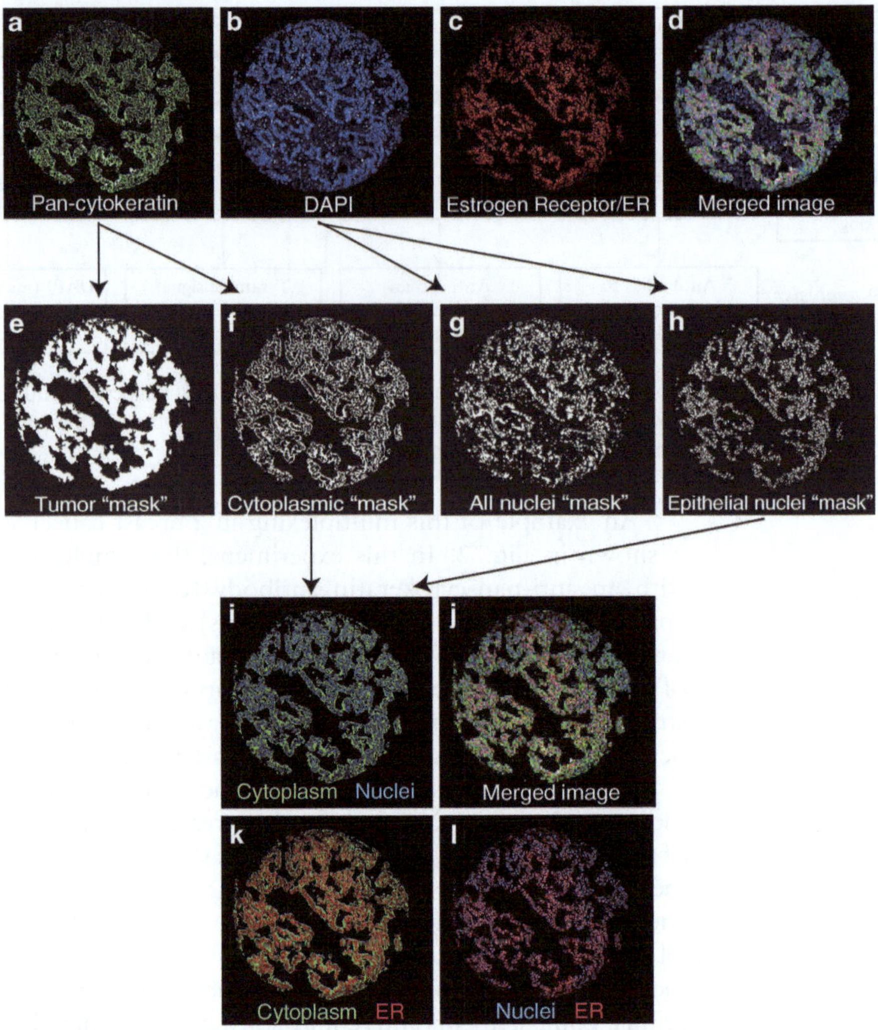

Fig. 2. Step-by-step illustration of AQUA® analysis for the quantification of estrogen receptor staining in epithelial nuclei of a breast cancer tissue microarray core. This schematic diagram illustrates the main concepts of AQUA analysis for the identification and masking of epithelial cells, cytoplasm and nuclei for the localization and measurement of the target (estrogen receptor) within those compartments.

2. Materials

2.1. Samples

1. Formalin-fixed, paraffin-embedded tissue samples cut to 5 μM thick and mounted on microscope slide.
2. Samples can be either whole tissue sections or prepared tissue microarrays.

2.2. Fluorescent Immunohistochemical Staining

1. Slides are initially blocked for nonspecific interaction and removal of endogenous peroxidase using Backround Sniper (Biocare, Concord, CA) and Peroxidazed (Biocare, Concord, CA).

2. Primary antibodies can be sourced from a broad range of vendors; however, the pan-cytokeratin primary antibody must be of a different species in order to prevent cross-reactivity.
3. The pan-cytokeratin antibody is probed with an Alexa555 dye-labeled secondary antibody (Molecular Probes, Eugene, OR). The primary antibody is targeted with Envision reagents (Dako, Carpenteria, CA), which provides multiple HRP moieties.
4. Primary biomarker amplification and visualization is further accomplished using a Cy5 conjugated tyramide signal amplification (TSA) system (Perkin Elmer, Waltham, MA.
5. Nuclei are visualized via counterstaining with DAPI, which is a component of the coverslip mounting media, Prolong Gold (Invitrogen, Carlsabad, CA).

2.3. Stained Slide Imaging

1. Slides are visualized on a PM-2000™ system, available from HistoRx, Inc. (New Haven, CT).

3. Methods

3.1. IF Staining (Further Detailed in (26))

An example of a stained breast cancer tissue microarray core is utilized for the description of AQUA® analysis below, including staining with:

1. Anti-pan-cytokeratin (Fig. 2a) to molecularly identify the epithelial regions and non-nuclear regions.
2. DAPI (Fig. 2b) to molecularly identify all nuclei.
3. Estrogen Receptor alpha (Fig. 2c) as the biomarker of interest.
4. The merged image is shown in Fig. 2d.

3.2. Image Collection

1. Images collected (such as by the PM2000™ platform (HistoRx, New Haven CT) or other output platform meeting the image specifications for AQUA® analysis software.

3.3. AQUA Analysis

The images collected contain all of the information necessary for determining cell types of interest (epithelial cells in this example), and within those regions subcellular compartments of interest (nuclear and cytoplasmic compartments in this example). AQUA analysis algorithms are applied for:

1. Objective quantification of biomarker expression on a continuous scale.
2. Localization of biomarkers within tissue, cell type(s) and/or compartment(s).
3. Standardization of results.

3.4. Epithelial Region Mask and Cytoplasmic Compartment Determinations

The pan-cytokeratin image (Fig. 2a) is thresholded (binarized) to generate

1. An epithelial cell mask (Fig. 2e) that excludes stromal regions.
2. A non-nuclear region mask (Fig. 2f).
3. All white pixels are then part of the mask, and all black pixels are not part of each of those masks.

3.5. Nuclear Compartment Determination

The DAPI image (Fig. 2b) is thresholded (binarized) to generate:

1. A mask of all nuclei within the sample (Fig. 2g) by subtracting out overlapping pixels with the cytoplasmic mask (Fig. 2f).
2. All white pixels are then a part of the nuclear mask, and all black pixels are not.
3. The nuclear mask (Fig. 2g) can be combined with the information from the epithelial mask (Fig. 2e) to generate a mask of only epithelial nuclei (Fig. 2h).

3.6. Visual Map of Pixel Designations

1. The composite pixel designations of the masks determined above are illustrated in 2I, with the nuclear compartment shown in blue and the cytoplasmic compartment shown in green (for epithelial regions).
2. The visual overlap of estrogen receptor and the pixel designations are shown in Fig. 2j.

3.7. Estrogen Receptor Quantification

The intensity of estrogen receptor staining (Fig. 2c) is then quantified in:

1. The epithelial cytoplasmic mask (overlap shown in Fig. 2k).
2. The epithelial nuclear mask (overlap shown in Fig. 2l with strong co-localization resulting in a purple color).
3. The resultant AQUA scores are then determined by the intensity of estrogen receptor within the nuclear or cytoplasmic masks, and then normalized for the area covered by the compartment of interest and for exposure time:

$$\text{AQUA scores} = AQ = \left(\frac{1}{\sum C_i}\right)\left(\sum T_i C_i\right)$$

Where $T_i = ((\text{Pixel intensity}/256)/\text{exposure time})$; and $C_i = 0$ or 1

4. Automated validation of all images by Validator™ is used to redact out samples with (1) Low mask intensity, (2) Low tissue presence, (3) High pixel saturation, and (4) Split TMA spots.

3.8. Standardization of Immunostaining Results and Utility with AQUA® Analysis

Even with AQUA analysis, can IHC staining of biomarkers on TMAs be truly quantitative? Several studies have demonstrated that, when carefully controlled, AQUA analysis can determine biomarker expression levels on a "molecules per cell" scale (14, 19) akin to the results one would achieve using an ELISA. Quantitative analysis of TMAs is dependent upon two factors (1) the use of internal standards for staining intensity and (2) the use of normalization factors to ensure that AQUA scoring is the same across staining runs, operators, and machines.

The first of these factors can be achieved by using cell line controls. Pellets composed of cell lines imbedded in agarose can be processed and included alongside tissues in TMAs. Since the level of protein expression in cell lysates can be assessed using a traditional ELISA, cell lines expressing a wide range of biomarker expression can be included in a TMA and serve as a standard "dilution series" control. This type of analysis has permitted the quantification of such markers as HER2/neu and beta-catenin (14, 19, 27).

The second is that the AQUA platform utilizes instrument calibration metholodogies that enable captured signal to be normalized across multiple instruments. Because AQUA technology is objective and strictly quantitative; it can be standardized across instruments (i.e., laboratories) as well as operators (i.e., observers). This is accomplished through a combination of light source and intrinsic machine calibration methodologies. The AQUA platform also employs software algorithmic methodologies that predominantly remove operator decisions from the image acquisition and scoring process (i.e., auto-exposure). These methodologies can be applied to any target of interest on any given tissue type to achieve percent coefficients of variation of less than 5%, which rival other quantitative immunoassays (i.e., ELISA and flow cytometry). These standardization methodologies have demonstrated for numerous markers, such as HER2 (27).

The AQUA analysis method thereby solves the problem of lack of reproducibility in IHC. In addition, it allows for more sophisticated statistical analyses of protein expression by utilizing the continuous, linear AQUA scores from TMA analysis as inputs for statistical algorithms. This is not possible with discreet and subjective IHC scores. By generating quantitative, continuous data for an immunoassay on tissue in a reproducible and standardized manner, AQUA scores can be utilized with confidence in many statistical analysis methodologies typically not applied to IHC results. Examples of such analyses are illustrated in Fig. 3, such as the utilization of AQUA scores for heatmaps (Fig. 3c) (28, 29), division of patient cohorts into groups based on continuous AQUA score data through the use of cluster analysis (*3D*), (30)), genetic algorithms (28), as well as for quantitative assessments of biomarker correlations (16) and associations (30)

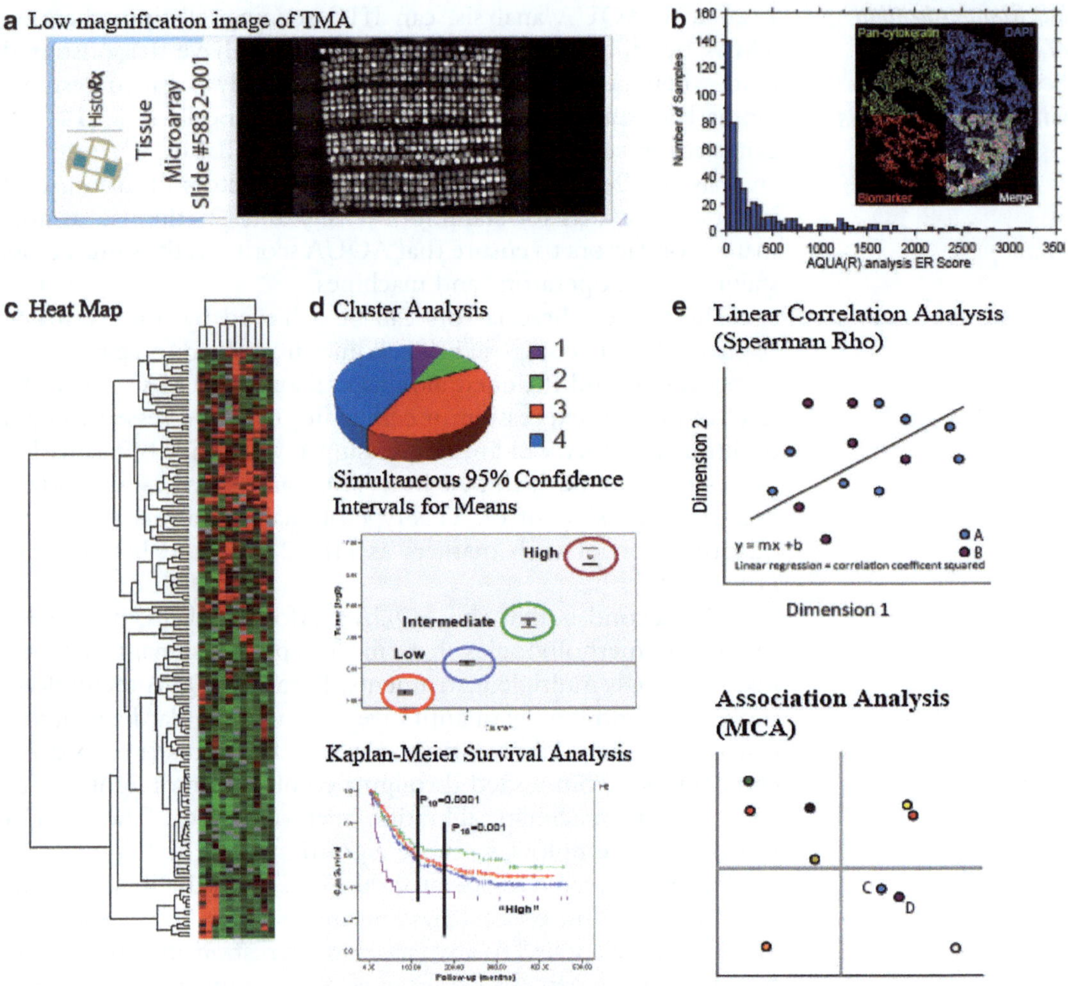

Fig. 3. Example types of analyses possible utilizing tissue microarray-based quantitative analysis from the generation of AQUA® scores. A low magnification image of a tissue microarray is shown in (**a**), with an example output of continuous AQUA score data from a TMA cohort shown in (**b**). Among the variety of different analyses that can be done with AQUA scores are the generation of heat maps (**c**), unsupervised clustering analyses for association with patient outcome such as Kaplan-Meier analyses (**d**), as well as correlation and association analyses (**e**) for assessment of biomarker relationship to patient outcome and drug response.

(Fig. 3e). Finally, AQUA analysis can be applied to the analysis of whole tissue section for the quantitative analysis of biomarker expression levels and heterogeneity (31).

3.9. Conclusions

The use of automated analysis of tissue microarrays by AQUA provides solutions to several key problems in the evaluation of protein expression – the need for a rapid method to score immunostaining, the need for objective analysis of protein expression levels and localization within samples, and the need for a reproducible way to solve the first two objectives in order to get the same answer every time a sample is evaluated.

4. Notes

To facilitate using AQUA® analysis methods in the laboratory, this section provides several key points which users should consider when working with the technology.

1. To reduce variations in quantitative measurements, strict process control to maintain reproducibility must be adopted early in the process – primarily with regards to the antigen retrieval and staining steps. This also includes the use of automation (i.e., automated staining systems).
2. The presence of multiple immunocomponents in the assay means that users should pay close attention to reagent qualification and quality control methods early in the development process. Reagents should be tested both individually and in combination with other assay components. Additionally, users should be aware that the electronic measurement of fluorescence images can be more sensitive than the human eye. Thus titers are typically more dilute for AQUA analysis than for chromagen IHC staining. Users need to pay particular attention to the background contributions components that may make subsequent analysis, which requires isolation of signal from background, difficult.
3. Tissue is a very complex matrix, and thus researchers using AQUA analysis need to use care when setting thresholds in the image analysis software.
4. AQUA scores can cover a wide range of values (4 logs) and the nature of fluorescence measurement is such that measurement variation increases with expression measurement. Thus, researchers typically use transformations, such as a log2, to normalize data sets.

Acknowledgments

RLC is supported by grants from the NIH/NCI including R21 CA 125277 and R21 CA 116265. DLR is supported by grants from the NIH including RO-1 CA 114277, R33 CA 106709 and R33 CA 110511.

References

1. Camp, R.L., L.A. Charette, and D.L. Rimm (2000) *Lab Invest.* **80**(12) 1943–9.
2. Wright, J.R., Jr. (1985) *Bull Hist Med.* **59**(3) 295–326.
3. Fowler, C.B., T.J. O'Leary, and J.T. Mason (2008) *Lab Invest.* **88**(7) 785–91.
4. Rimm, D.L., J.M. Giltnane, C. Moeder, M. Harigopal, G.G. Chung, R.L. Camp, and B. Burtness (2007) *J Clin Oncol.* **25**(17) 2487–8.
5. Rimm, D.L. (2006) *Nat Biotechnol.* **24**(8) 914–6.

6. Camp, R.L., G.G. Chung, and D.L. Rimm (2002) *Nat Med.* **8**(11) 1323–7.
7. Rao, J., D. Seligson, and G.P. Hemstreet (2002) *Biotechniques.* **32**(4) 924–6, 928–30, 932.
8. Giltnane, J.M., L. Ryden, M. Cregger, P.O. Bendahl, K. Jirstrom, and D.L. Rimm (2007) *J Clin Oncol.* **25**(21) 3007–14.
9. Camp, R.L., M. Dolled-Filhart, B.L. King, and D.L. Rimm (2003) *Cancer Res.* **63**(7) 1445–8.
10. Wolff, A.C., M.E. Hammond, J.N. Schwartz, K.L. Hagerty, D.C. Allred, R.J. Cote, M. Dowsett, P.L. Fitzgibbons, W.M. Hanna, A. Langer, L.M. McShane, S. Paik, M.D. Pegram, E.A. Perez, M.F. Press, A. Rhodes, C. Sturgeon, S.E. Taube, R. Tubbs, G.H. Vance, M. van de Vijver, T.M. Wheeler, and D.F. Hayes (2007) *J Clin Oncol.* **25**(1) 118–45.
11. Vani, K., S.R. Sompuram, P. Fitzgibbons, and S.A. Bogen (2008) *Arch Pathol Lab Med.* **132**(2) 211–6.
12. Allred, D.C., J.M. Harvey, M. Berardo, and G.M. Clark (1998) *Mod Pathol.* **11**(2) 155–68.
13. McCarty, K.S., Jr., E. Szabo, J.L. Flowers, E.B. Cox, G.S. Leight, L. Miller, J. Konrath, J.T. Soper, D.A. Budwit, W.T. Creasman, and et al. (1986) *Cancer Res.* **46**(8 Suppl) 4244s–4248s.
14. McCabe, A., M. Dolled-Filhart, R.L. Camp, and D.L. Rimm (2005) *J Natl Cancer Inst.* **97**(24) 1808–15.
15. Zheng, Z., T. Chen, X. Li, E. Haura, A. Sharma, and G. Bepler (2007) *N Engl J Med.* **356**(8) 800–8.
16. Pozner-Moulis, S., M. Cregger, R.L. Camp, and D.L. Rimm (2007) *Lab Invest.* **87**(3) 251–60.
17. Garraway, L.A., H.R. Widlund, M.A. Rubin, G. Getz, A.J. Berger, S. Ramaswamy, R. Beroukhim, D.A. Milner, S.R. Granter, J. Du, C. Lee, S.N. Wagner, C. Li, T.R. Golub, D.L. Rimm, M.L. Meyerson, D.E. Fisher, and W.R. Sellers (2005) *Nature.* **436**(7047) 117–22.
18. Psyrri, A., Z. Yu, P.M. Weinberger, C. Sasaki, B. Haffty, R. Camp, D. Rimm, and B.A. Burtness (2005) *Clin Cancer Res.* **11**(16) 5856–62.
19. Rubin, M.A., M.P. Zerkowski, R.L. Camp, R. Kuefer, M.D. Hofer, A.M. Chinnaiyan, and D.L. Rimm (2004) *Am J Pathol.* **164**(3) 831–40.
20. Dolled-Filhart, M., A. McCabe, J. Giltnane, M. Cregger, R.L. Camp, and D.L. Rimm (2006) *Cancer Res.* **66**(10) 5487–94.
21. Berger, A.J., D.W. Davis, C. Tellez, V.G. Prieto, J.E. Gershenwald, M.M. Johnson, D.L. Rimm, and M. Bar-Eli (2005) *Cancer Res.* **65**(23) 11185–92.
22. Psyrri, A., A. Bamias, Z. Yu, P.M. Weinberger, M. Kassar, S. Markakis, D. Kowalski, E. Efstathiou, R.L. Camp, D.L. Rimm, and M.A. Dimopoulos (2005) *Clin Cancer Res.* **11**(23) 8384–90.
23. Moeder, C.B., J.M. Giltnane, S.P. Moulis, and D.L. Rimm (2009) *Methods Mol Biol.* **520**:163–75.
24. Kluger, H.M., S.F. Siddiqui, C. Angeletti, M. Sznol, W.K. Kelly, A.M. Molinaro, and R.L. Camp (2008) *Lab Invest.* **88**(9) 962–72.
25. Ross, J.S., K. Mistry, K.B. Bacon, and R.D. Camp (1991) *J Immunol Methods.* **140**(2) 219–25.
26. Moeder, C., J. Giltnane, S. Pozner-Moulis, and D.L. Rimm, *Quantitative, Fluorescence-based In-Situ Assessment of Protein Expression.* Methods in Molecular Biology: Tumor Marker Discovery. in press, Totowa New Jersey: Humana Press.
27. Gustavson, M.D., B. Bourke-Martin, D.M. Reilly, M. Cregger, C. Williams, J. Mayotte, M. Zerkowski, G. Tedeschi, R. Pinard, and J. Christiansen (2009) *Arch Pathol Lab Med.* **133**:1413–19.
28. Dolled-Filhart, M., L. Ryden, M. Cregger, K. Jirstrom, M. Harigopal, R.L. Camp, and D.L. Rimm (2006) *Clin Cancer Res.* **12**(21) 6459–68.
29. Siddiqui, S.F., J. Pawelek, T. Handerson, C.Y. Lin, R.B. Dickson, D.L. Rimm, and R.L. Camp (2005) *Cancer Epidemiol Biomarkers Prev.* **14**(11 Pt 1) 2517–23.
30. Dolled-Filhart, M., R. Pinard, D. Waldron, A. Ang, L. Goodrich, S. Myrand, D. Thornton, J. Graff, and B. Mullaney. *Clustering of phosphoproteins targeted by enzastaurin identifies significant biomarker association with patient outcome and novel associations between biomarker groupings in a glioblastoma multiforme cohort.* in *AACR Annual Meeting.* 2008. San Deigo, CA.
31. Zerkowski, M.P., R.L. Camp, B.A. Burtness, D.L. Rimm, and G.G. Chung (2007) *Cancer Invest.* **25**(1) 19–26.

Chapter 16

Digital Microscopy for Boosting Database Integration and Analysis in TMA Studies

Tibor Krenacs, Levente Ficsor, Sebestyen Viktor Varga, Vivien Angeli, and Bela Molnar

Abstract

The enormous amount of clinical, pathological, and staining data to be linked, analyzed, and correlated in a tissue microarray (TMA) project makes digital slides ideal to be integrated into TMA database systems. With the help of a computer and dedicated software tools, digital slides offer dynamic access to microscopic information at any magnification with easy navigation, annotation, measurement, and archiving features. Advanced slide scanners work both in transmitted light and fluorescent modes to support biomarker testing with immunohistochemistry, immunofluorescence or fluorescence in situ hybridization (FISH). Currently, computer-driven integrated systems are available for creating TMAs, digitalizing TMA slides, linking sample and staining data, and analyzing their results. Digital signals permit image segmentation along color, intensity, and size for automated object quantification where digital slides offer superior imaging features and batch processing. In this chapter, the workflow and the advantages of digital TMA projects are demonstrated through the project-based MIRAX system developed by 3DHISTECH and supported by Zeiss.

The enhanced features of digital slides compared with those of still images can boost integration and intelligence in TMA database management systems, offering essential support for high-throughput biomarker testing, for example, in tumor progression/prognosis, drug discovery, and target therapy research.

Key words: Digital microscopy, Tissue microarray, Database integration, Validated scoring, Image segmentation, Correlation analysis

1. Introduction

TMA studies may include thousands of stained samples to be analyzed and correlated with several clinico-pathological parameters multiplying the amount of information to be considered (1). Assessment of TMA staining results with manual microscopy is extremely tedious, time-consuming, and prone to misinterpretation

Ronald Simon (ed.), *Tissue Microarrays: Methods and Protocols*, Methods in Molecular Biology, vol. 664,
DOI 10.1007/978-1-60761-806-5_16, © Springer Science+Business Media, LLC 2010

due to the high risk of mixing up samples and the lack of proper validation. Therefore, the bottleneck of TMA studies for several years since its introduction in 1998 (2) had been the lack of database management tools offering safe and easy handling of alphanumerical and image data linked together.

Digital imaging technologies offer output formats, which can be easily integrated into computer databases and used for automated image analysis (3). However, digital still images represent only a fragment of normal-sized slides and do not allow the study of fine details even in spots as small as 0.6 mm in diameter. On the contrary of this, digital slides represent whole microscopic slides assembled from many still images, where low power views are gained with increasing compression of the original tiled images. This feature allows dynamic and immediate access to in focus images of any part of the slide at any microscopic magnification with superior details in the context of the whole slide traced on simultaneously running previews. Permanent batch linking of stained TMA spots with any sample data, batch processing during analysis, easy annotations, measurements, and archiving are among the enhanced features of digital slides compared to static images, which can boost fluency, precision, and consistency of TMA studies.

1.1. TMA Database Management Systems with Integrated Digital Imaging

Since database management starts at case selection, advanced systems must cover the whole TMA procedure from design to interpretation with the integration of digital H&E slides and images of donor paraffin blocks used at array building and digital slides of stained TMAs for analysis. However, modern digital slide technology and the supporting informatics, including high speed computer processors and mass storage devices, have become available only in recent years, so some integrated TMA database systems still utilize static digital images (3).

Basically, two classes of TMA project management systems utilizing digital imaging have become available recently. Open-access web-based systems such as TMAD (4), TMAJ (5) and TAMME (6) (see Note 1) offer software tools for examining TMA results and focus on building large integrated TMA data repositories available for the wide research community. Their advantage is the use of standard ontologies, classifications, and data representations intended to be compatible with Tissue Microarray Data Exchange Specifications (7, 8), which permit the merging of TMA data of different laboratories and expanding existing projects with new data classes (9). However, they utilize only static images, hindering users' immediate access to fine and sometimes critical details.

The commercial systems either offer their own TMA arrayer and/or scanner besides software tools for supporting the TMA workflow or borrow instruments and mainly focus on the analysis of TMA results with their software tools. Some commercial TMA management systems are also restricted to the use of digital still images such as the SpotBroser/Tysalis (10) and TMAx (11, 12). More advanced systems utilize digital TMA slides such as the MIRAX (13, 14), TMA LabII (15), Ariol (16) and TMAscore (17) (see Note 2). However, neither the software nor the digital slide formats of these systems are compatible with each other to facilitate data exchange and building open access cross-platform TMA databases. Nevertheless, all these major systems offer software tools for image analysis and quantification using selected fields of digital images.

1.2. Efficiency of TMA Database Systems

Factors critical to the efficiency of TMA database manager systems are (1) the software's compatibility with input data formats; (2) the level of data integration, and ease of accessing images and any sample data linked during assessment; (3) the software's ability to allow free sorting of stained spots, with any of their data revealed, into galleries independently of their original physical locations; (4) the availability of image quantification software running on the same platform and allowing batch processing; and (5) the output files' compatibility with those used in software of relation/cluster, survival, and statistical analysis.

It is also obvious that hardware and software tools offered by the same developer for the whole TMA process improve the chances of workflow integration and system intelligence compared to solutions combining hardware and software tools from different sources. Concerning efficiency and system integration, there can be significant differences between the listed TMA data management systems which, however, are out of the focus of this chapter. Here we concentrate on introducing one of the most integrated systems – the MIRAX, including a TMA arrayer and a family of scanners armed with software tools for the management of complete TMA studies from design to interpretation. The MIRAX TMA module software supports project-based handling of many TMA slides and their related data at the same time, in contrast to the simple slide-based systems that work with only one slide at a time.

2. Materials

The role of digital slides in MIRAX TMA projects is demonstrated in the context of the whole process from design to analysis. Apart from the standard laboratory tools and disposables, all tools mentioned in the list below are included in the respective instrument setups.

1. Digital Excel spreadsheet including all data (clinical, pathological, epidemiological and histotechnical) (Fig. 1) of

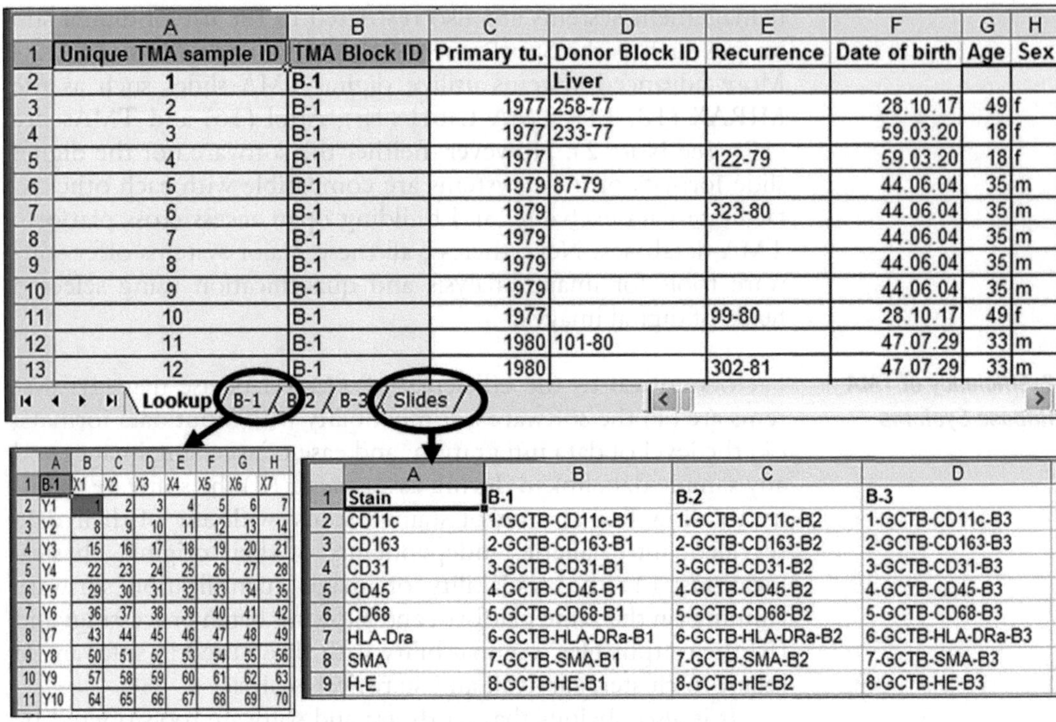

	A	B	C	D	E	F	G	H
1	Unique TMA sample ID	TMA Block ID	Primary tu.	Donor Block ID	Recurrence	Date of birth	Age	Sex
2	1	B-1		Liver				
3	2	B-1	1977	258-77		28.10.17	49	f
4	3	B-1	1977	233-77		59.03.20	18	f
5	4	B-1	1977		122-79	59.03.20	18	f
6	5	B-1	1979	87-79		44.06.04	35	m
7	6	B-1	1979		323-80	44.06.04	35	m
8	7	B-1	1979			44.06.04	35	m
9	8	B-1	1979			44.06.04	35	m
10	9	B-1	1979			44.06.04	35	m
11	10	B-1	1977		99-80	28.10.17	49	f
12	11	B-1	1980	101-80		47.07.29	33	m
13	12	B-1	1980		302-81	47.07.29	33	m

	A	B	C	D	E	F	G	H
1	B-1	X1	X2	X3	X4	X5	X6	X7
2	Y1	1	2	3	4	5	6	7
3	Y2	8	9	10	11	12	13	14
4	Y3	15	16	17	18	19	20	21
5	Y4	22	23	24	25	26	27	28
6	Y5	29	30	31	32	33	34	35
7	Y6	36	37	38	39	40	41	42
8	Y7	43	44	45	46	47	48	49
9	Y8	50	51	52	53	54	55	56
10	Y9	57	58	59	60	61	62	63
11	Y10	64	65	66	67	68	69	70

	A	B	C	D
1	Stain	B-1	B-2	B-3
2	CD11c	1-GCTB-CD11c-B1	1-GCTB-CD11c-B2	1-GCTB-CD11c-B3
3	CD163	2-GCTB-CD163-B1	2-GCTB-CD163-B2	2-GCTB-CD163-B3
4	CD31	3-GCTB-CD31-B1	3-GCTB-CD31-B2	3-GCTB-CD31-B3
5	CD45	4-GCTB-CD45-B1	4-GCTB-CD45-B2	4-GCTB-CD45-B3
6	CD68	5-GCTB-CD68-B1	5-GCTB-CD68-B2	5-GCTB-CD68-B3
7	HLA-Dra	6-GCTB-HLA-DRa-B1	6-GCTB-HLA-DRa-B2	6-GCTB-HLA-DRa-B3
8	SMA	7-GCTB-SMA-B1	7-GCTB-SMA-B2	7-GCTB-SMA-B3
9	H-E	8-GCTB-HE-B1	8-GCTB-HE-B2	8-GCTB-HE-B3

Fig. 1. Excel database for digital TMA management consists of three groups of worksheets. A "Lookup" table containing clinico-pathological data of the samples. TMA maps for each TMA block (B-1, B-2…B-n) and a "Slides" sheet linking stains and digital slide names to TMA blocks. Serial slides within a block are numbered consecutively, for example, 1–8 to support core reconstruction (*bottom right panel*). Columns A and B in the Lookup and all TMA maps are automatically generated at array building with the TMA Master.

cases selected to fit the study design, in the format required for both the TMA Master and TMA Module software (see Note 3).

2. Donor tissue blocks with their H&E-stained digital slides (see *scanning later and* Note 4).

3. TMA Master instrument (Fig. 2), including a camera, external light source, computer and arrayer software (version 2.0) for computer controlled creation of recipient and TMA blocks (see Note 5).

4. Row paraffin blocks made in standard metal molds of 6 mm depth using standard plastic cassettes (28 × 40 × 6 mm) for making recipient blocks (see Note 6).

5. Stained TMA sections mounted on standard 25 × 75 × 1.0 mm glass slides (see Note 7), which are labeled consecutively either with barcode or manually. Label should include information on project and block ID, date and type of staining, and serial number of slides within the same block.

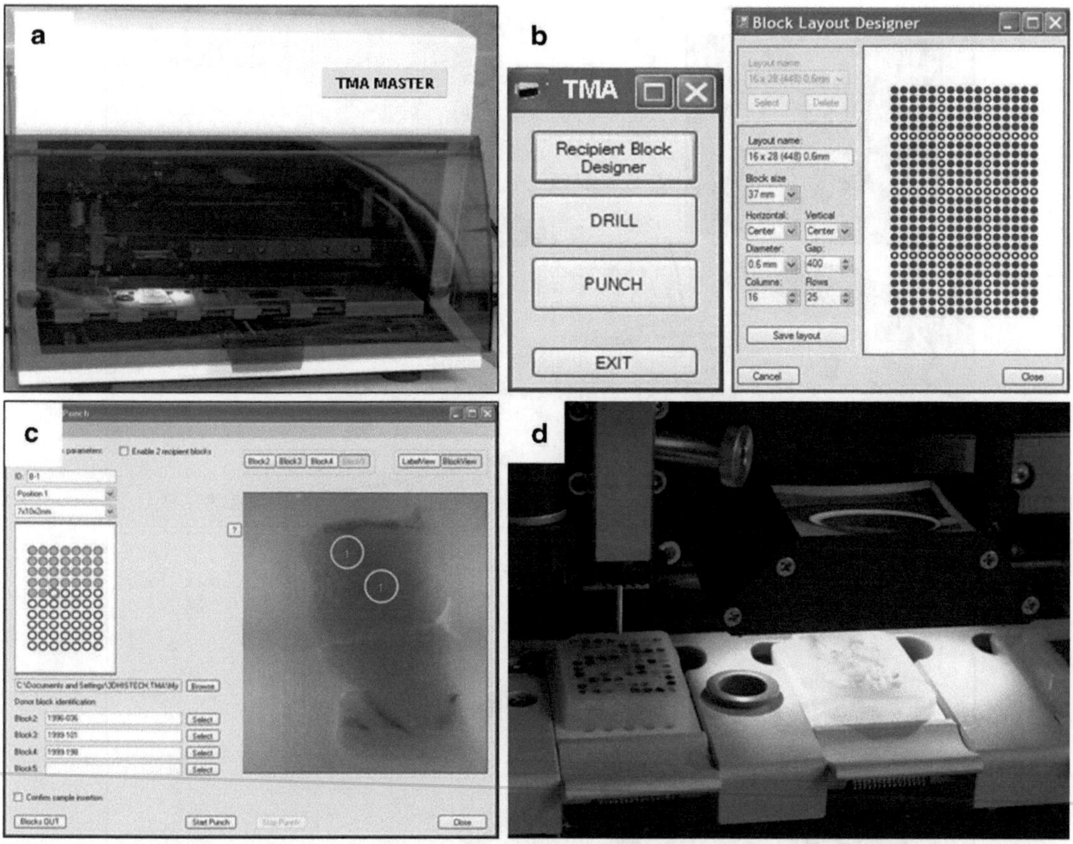

Fig. 2. Steps of TMA array building with the TMA Master (**a**), housing five blocks at one time, and its software involve: (**b**) Designing array formats and drilling holes in the required number of recipient blocks accordingly; (**c**) Importing Excel Lookup data and array building by punching out donor tissue areas (*circles*) predefined on digital H&E slides based on their alignment with the high resolution image of donor blocks. (**d**) Close-up view of array building in the instrument.

6. Digital slide scanner MIRAX Scan with a vertical slide rack (see Note 8) (Fig. 3) linked to a driving computer (software version 1.10.23) and light source either for transmitted light and/or fluorescent scanning, the latter is supported with proper filter set inserted into the scanners rotary wheel.
7. Internal or external hard disk drivers with a substantial storage capacity of tens of GB (see Note 9) for each project, or a server, either internal or external.
8. Digital slide viewer software (MIRAX Viewer version) platform with TMA Module and HistoQuant software (version 1.10.13) for database integration, management, scoring, and quantification of TMA slides (see Note 10).
9. Software tools for survival, cluster, and statistical analysis (see Note 11).

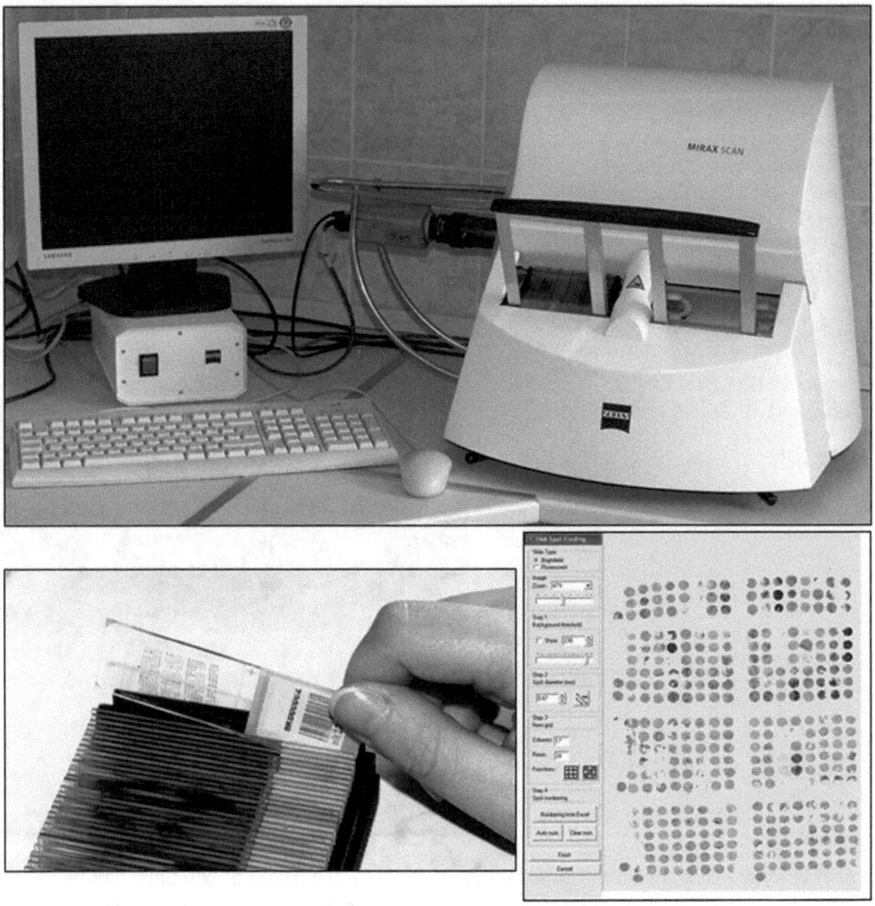

Fig. 3. MIRAX Scan digital microscope for scanning H&E-stained slides of donor blocks and the stained TMA slides (*right bottom*), fed in vertical slide racks of 50-slide capacity (*left bottom*). MIRAX Scan can take six of these racks allowing feeding of 300 glass slides in one run.

3. Methods

The workflow of MIRAX digital TMA management system described below is summarized in Fig. 4.

3.1. Designing, Creation, and Data Integration of TMAs Using the TMA Master

1. All data classes needed for sample identification and correlation with the staining are collected in an Excel sheet from patients' samples meeting the study design (Fig. 1), for example, normal tissues, pre-malignant, malignant and metastatic lesions from a given tissue/tumor type for a tumor progression assay (1).
2. H&E-stained slides of the selected donor blocks are scanned into digital slides using the MIRAX Scan (Fig. 3), and representative areas for punching are labeled digitally using the MIRAX Viewer (see Note 4).

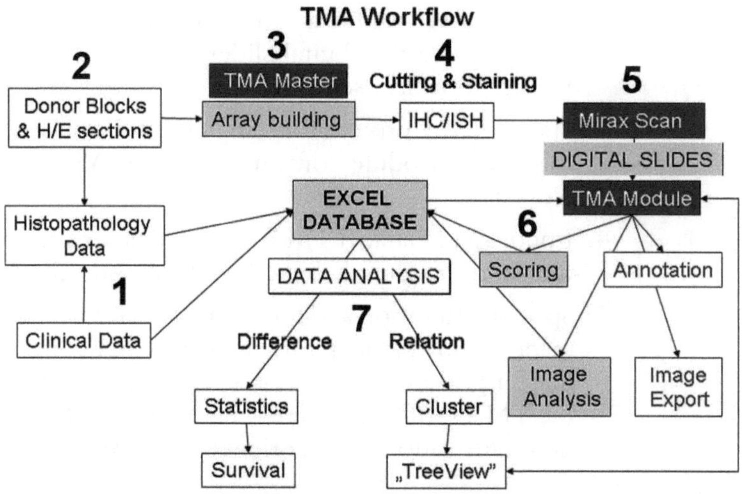

Fig. 4. Workflow of a Digital TMA study using the MIRAX system: (1) Collecting in Excel of all data of patients/samples to be correlated with staining; (2) Sorting donor blocks with their respective H&E-stained digital slides; (3) Designing and creating TMA recipient blocks and array building using the TMA Master; (4) Cutting TMA slides and standard staining of target biomarkers using IHC or ISH; (5) Digitalizing stained TMA slides using the MIRAX Scan; (6) Running a "TMA project" on relevant digital TMA slides including (**a**) importing sample related Excel database, (**b**) spot identification, (**c**) setting up the scoring scheme(s) for studied markers, (**d**) validated on-screen scoring of digital slide spots, (**e**) image analysis/ quantification using the HistoQuant software, and (**f**) exporting database including input data added with the scores/ quantification results; (7) Testing results with statistics, survival and/or cluster analysis using compatible software tools.

3. Donor blocks are sorted in increasing numerical order of their ID numbers and years (see Note 4).
4. TMA recipient blocks are designed (see Note 12) according to the number of cases and the need for replica cores in the study, and drilled with the TMA Master instrument under computer control (Fig. 2).
5. Recipient and donor blocks are clipped into TMA Master (see Note 6) and XLS sample data imported. Areas pre-selected on digital H&E slides are labeled on high resolution images of the donor blocks for punching. An orientation core is used (see Note 13).
6. Clicking on the donor sample ID of the imported XLS list and indicating its position in the recipient block drives punching and links core and patient data. A TMA map is automatically generated accordingly (see Note 3).
7. Saving XLS database will include a Lookup table and the relevant TMA map. Saving session parameters allow one to continue a project, for example, on the next day if it is interrupted.

3.2. Digitalization of TMA Slides and Setting Up Projects

1. TMA blocks are cut sections and stained according to standard and automated procedures, for example, for immunostaining or in situ hybridization either chromogenic or fluorescent (see Note 7).

2. Stained TMA slides are digitalized using the MIRAX Scan (Fig. 2), and the digital slides are opened with the MIRAX Viewer software (see Note 9).
3. Start a TMA project for parallel multi-user assessment using the TMA Module software from the Viewer platform (see Note 10).
4. Import XLS data. If TMA Master was used for array building, the cores are already linked to their respective data.
5. Spot identification and linking image and sample data of TMAs made on any platform other than MIRAX. By using the TMA map, a grid of square bookmarks is pulled over and linked to the TMA spots, which are then automatically numbered based on their relation to the orientation core (see Note 3).

3.3. Scoring and Image Analysis in Digital TMA Projects

1. Setting up arbitrary and separate scoring schemes for biomarkers even within the same project considering frequency and intensity of the staining.
2. Scoring of identified spots on digital slides from the Viewer function by calling in relevant scoring scheme (see Note 14) with the right mouse button and selecting the appropriate category of score which is immediately linked to the spots' bookmark.
3. Alternative on-screen scoring from the Gallery collections of stained spots, where any data set available from imported XLS can be revealed beside the given spot during scoring. All these features are used as a single or combined filter option for free sorting of spots independently of their physical location (Fig. 5).
4. Validation of scores by using the filter sorting and direct comparison of spots, for example, those initially thought to be of the same score class for reconsideration.
5. Digital annotation of representative sample areas on a series of spots for batch quantification of image objects and stained areas (Fig. 6) by using the HistoQuant software module opened from the Viewer/TMA platform (see Note 15).
6. Image object specific parameters such as perimeter, area, average intensity, number, percentage of positive cells, etc., and area specific parameters such as selected area, positive area, percentage of positive area, average intensity of positive area, etc., are measured and scatter plots are created immediately (Fig. 6).
7. Exporting annotated images or arbitrary image gallery collections in any standard image formats including TIFF, JPEG, or HTML for reporting or publication.
8. Exporting analysis and scoring results in different formats including CSV, XLS, and Cluster, compatible with most software sets used for cluster, statistics, and survival analysis (see Note 11).

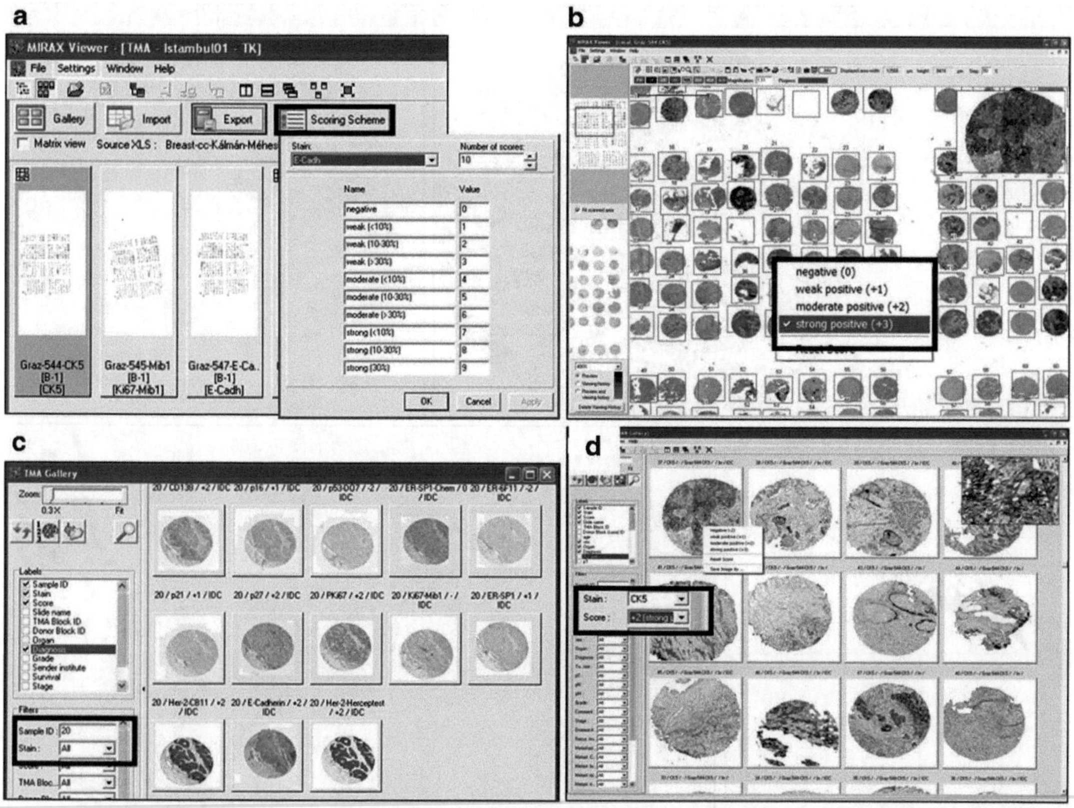

Fig. 5. Scoring a digital slide project after Excel database import: (**a**) Setting up arbitrary scoring scheme considering both intensity and frequency of stained cells; (**b**) Selecting score from a scroll-down menu for each identified spot using the Viewer function. Filtering options in the Gallery support: (**c**) Easy assessment of expression profiles and validation of spot identification by arranging side-by-side the serial slides of the same core stained differently. (**d**) Validation of scoring allowing reassessment of different spots achieving the same score for a marker staining.

4. Notes

1. Examples of major open-access web-based systems offering digital image repositories are the Tissue Microarray Database (TMAD) by Stanford University (Stanford, CA, USA; http://tma.stanford.edu); TMAJ by John Hopkins University (Baltimore, MD, USA; http://tmaj.pathology.jhmi.edu), or the TAMME by Graz University (Austria; http://genome.tugraz.at/Software/TAME). The Human Proteome Atlas (HPA; http://www.proteinatlas.org) project (Uppsala and Stockholm, Sweden) generating their own data for public access also belong here (11).
2. Specification of some of the commercial TMA management systems: MIRAX TMA Module by 3DHISTECH Ltd (Budapest, Hungary and Zeiss GmbH, Germany) (13, 14); TMA LabII by Aperio Inc (Vista, CA, USA) (15); Ariol

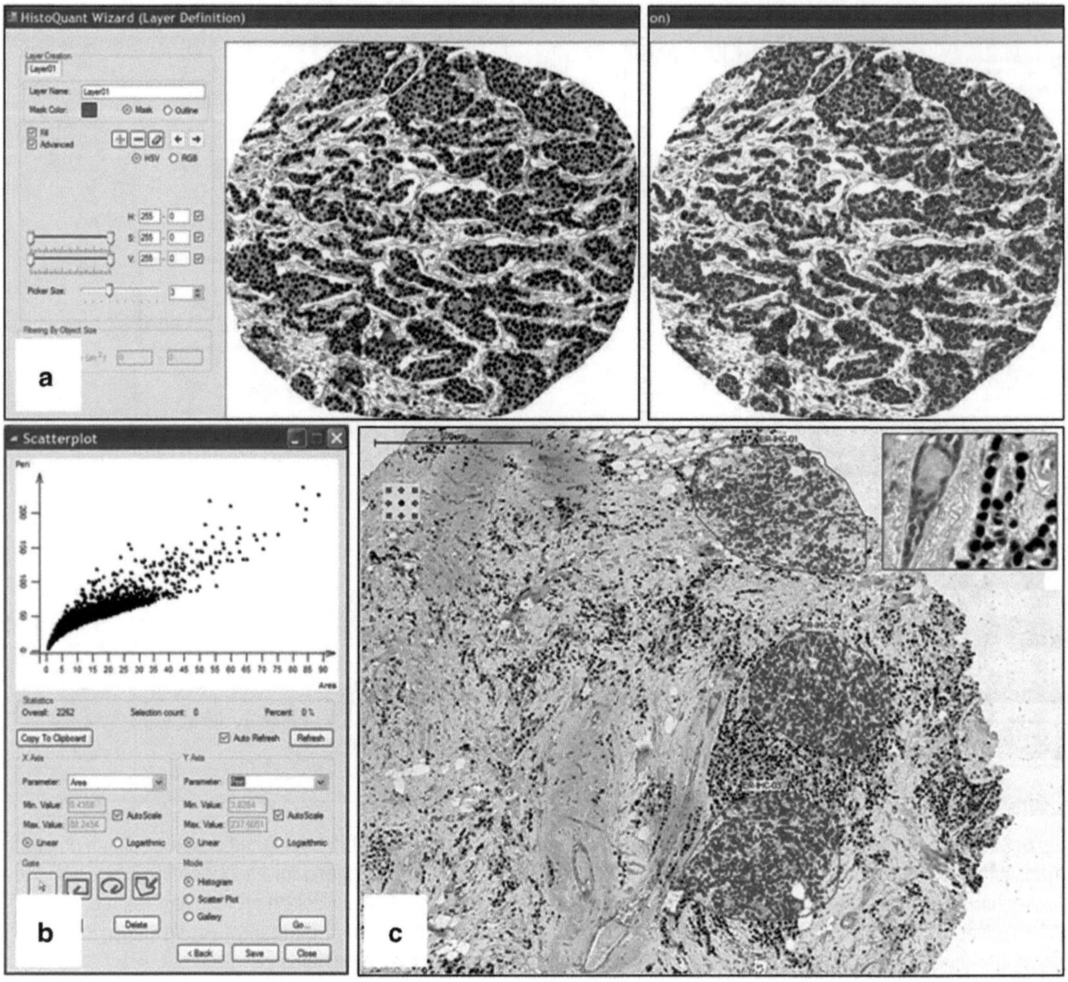

Fig. 6. Image analysis using the HistoQuant software. Area of interest is selected with annotation on a digital slide and imported for analysis. (**a**) MIRAX image segmentation profile (.misp) based on color, intensity, and size thresholds is created highlighting signals (*red mask*) above the threshold for quantification, for example, of estrogen receptor (ER) positive cell number and positive object (*nuclei*) parameters (area, perimeter, average intensity, etc.) to be demonstrated on scatter plots (**b**). (**c**) The .misp files are used for batch processing of arbitrary number of areas through the same analysis profile to be relocalized again on digital slides.

(formerly by Applied Imaging) (16) by Genetix Ltd (New Milton, England); TMAscore by Bacus Inc (Lombard. IL, USA) (17); SpotBroser/Tysalis by Alphelys (Plaisir, France) (10) and TMAx by Beechers Inc (Sun Prairie, WI, USA) (11, 12).

3. In the Excel Lookup table, each donor paraffin block and all related data are represented in one row and each column specifies a data class to be considered, for example, donor block ID, diagnosis, etc. (Fig. 1). If TMA Master is used for array building, a TMA map will be created in a separate worksheet within the same Excel file. Also, the extra columns of "Unique sample ID," containing numbers showing the

position of the given core in the TMA map, and of "Block-ID" are automatically generated. For TMAs built on a different platform, these columns and the TMA map for each block must be created manually. A third group of worksheet is needed for linking staining specification to digital slides.

4. Case numbers should be in a format where number and year are linked with a dash, for example, 199–07, both in Excel sheets and as starting figures for H&E slides at scanning to be easily sorted, that is, in an increasing numerical order to match donor block order.
5. The MIRAX TMA database management system is compatible with any TMA format created with any instrument. However, when TMA Master is used, an Excel database is automatically created including TMA map and Lookup table with core and sample data already linked to facilitate later analysis.
6. Standard plastic cassettes fit into the cassette holders of TMA Master. Recipient paraffin blocks can be of any sensible x–y dimensions, but it is advisable to have a standard thickness of 6 mm above the plastic.
7. Standard-sized adhesive glass slides such as SuperFrost Ultra/Plus (Menzel GmbH Braunschweig, Germany) are optimal for all staining procedures permitting loss of tissue spots and blockage-free feeding of slides at vertical scanning.
8. MIRAX Scan digitalizes glass slides in a vertical position. Each storage rack for vertical scanning can house 50 slides at the maximum, and six of these racks can be scanned with the instrument in one run making up 300 samples in one go. MIRAX Midi is a horizontal scanner using trays of 12 slides in one run, which can also work in fluorescent mode. MIRAX Desk is a horizontal scanner for single slide feeding, which does not support fluorescence.
9. By using the MIRAX Scan, digital TMA slides with x–y dimensions close to those of the plastic cassette usually require 0.6–1 GB storage space each with ~100 MB/min scanning time depending on the stained core/spot size.
10. Minimal system requirements for running MIRAX software include, operation system for all: Microsoft Windows XP Home, XP Professional 32-bit, or Vista 32-bit; TMA Master: Intel Pentium4 processor, 1 GB RAM; MIRAX Scan: Intel Pentium4 2.4 GHz (Core 2 Duo recommended), 3 GB RAM; and MIRAX Viewer including HistoQuant: Intel Pentium4 2.4 GHz, 1 GB RAM.
11. Examples of software sets for Cluster analysis: Cluster 2.11 and TreeView *1.60* (http://rana.lbl.gov/EisenSoftware.htm) (18, 19); Statistics: SPSS 13.0 (http://www.spss.com);

and Survival analysis: GraphPad Prism 4. (http://www.graphpad.com/prism).

12. Recipient paraffin blocks with x–y dimensions close to those of a standard plastic cassette (23 × 30 or 23 × 36 mm) permit drilling 448 holes of 0.6 mm, 240 holes of 1.0 mm and 80 holes of 2.0 mm at the maximum leaving at least 0.5 mm paraffin between cores and a 1.5 mm edge margin of paraffin intact.
13. A core of normal tissue (e.g., liver) easily differentiated from most test tissues should be inserted into the starting corner position of all TMA blocks ensuring safe TMA spot identification at any case of slide flipping or tilting at mounting. The edge of paraffin at this corner should be cut off before section cutting to support the mounting of TMA slides onto glass slides at standard orientation.
14. Scoring can be done on the monitor at any random order since each spot and score are directly linked to their bookmark ID number so as the given score.
15. Image segmentation on selected areas is based on object color, intensity and size. Analysis profile of a staining is set up on annotated areas representing the given tissue/tumor type with the creation of a "MIRAX image segmentation profile (.misp)" file, which is used for batch processing of a series of annotated sample areas (Fig. 6).

References

1. Kallioniemi, O-P., Wagner, U., Kononen, J., Sauter, G. (2001) Tissue microarray technology for high-throughput molecular profiling of cancer. Hum. Mol. Gen. 10:657–662.
2. Kononen, J., Bubendorf, L., Kallioniemi, A., Barlund, M., Schraml, P., Leighton, S., Torhorst, J., Mihatsch, M.J., Sauter, G., Kallioniemi, O.P. (1998) Tissue microarrays for high through-put molecular profiling of tumor specimens. Nat. Med. 4:844–847.
3. Kayser, K., Molnar, B., and Weinstein, R.S. (2006) Virtual slides technology. In: K. Kayser, B. Molnar, and R.S. Weinstein (eds.), Virtual microscopy: fundamentals, applications, perspectives of electronic tissue-based diagnosis. VSV Publlication, Berlin, pp. 103–123.
4. Marinelli, R.J., Montgomery, K., Liu, C.L., Shah, N.H., Prapong, W., Nitzberg, M., Zachariah, Z.K., Sherlock, G.J., Natkunam, Y., West, R.B., van de Rijn, M., Brown, P.O., Ball, C.A. (2008) The Stanford tissue microarray database. Nucleic Acids Res. 36:D871–D817
5. Faith, D.A., Isaacs, W.B., Morgan, J.D., Fedor, H.L., Hicks, J.L., Mangold, L.A., Walsh, P.C., Partin, A.W., Platz, E.A., Luo, J., De Marzo, A.M. (2004) Trefoil factor 3 overexpression in prostatic carcinoma: prognostic importance using tissue microarrays. Prostate. 61:215–227.
6. Thallinger, G.G., Baumgartner, K., Pirklbauer, M., Uray, M., Pauritsch, E., Mehes, G., Buck, C.R., Zatloukal, K., Trajanoski, Z. (2007) TAMEE: data management and analysis for tissue microarrays. BMC Bioinformatics. 8:81.
7. Kajdacsy-Balla, A., Geynisman, J.M., Macias, V., Setty, S., Nanaji, N.M., Berman, J.J., Dobbin, K., Melamed, J., Kong, X., Bosland, M., Orenstein, J., Bayerl, J., Becich, M.J., Dhir, R., Datta, M.W. (2007) Practical aspects of planning, building, and interpreting tissue microarrays: the Cooperative Prostate Cancer Tissue Resource experience. J. Mol. Histol. 38:113–121.
8. Lee, H.W., Park, Y.R., Sim, J., Park, R.W., Kim, W.H., Kim, J.H. (2006) The tissue microarray object model: a data model for storage, analysis, and exchange of tissue microarray experimental data. Arch. Pathol. Lab. Med.130:1004–1013.

9. Berman, J.J., Datta, M., Kajdacsy-Balla, A., Melamed, J., Orenstein, J., Dobbin, K., Patel, A., Dhir, R., Becich, M.J. (2004) The tissue microarray data exchange specification: implementation by the Cooperative Prostate Cancer Tissue Resource. BMC Bioinformatics. 27:5–19.
10. Galon, J., Costes, A., Sanchez-Cabo, F., Kirilovsky, A., Mlecnik, B., Lagorce-Page, C., Tosolini, M., Camus, M., Berger, A., Wind, P., Zinzindohoue, F., Bruneval, P., Cugnenc, P-H., Trajanoski, Z., Fridman, W-H., Page, F. (2006) Type, density, and location of immune cells within human colorectal tumors predict clinical outcome. Science 313:1960–1965.
11. Hober, S. and Uhlen, M. (2008) Human protein atlas and the use of microarray technologies. Curr. Opin. Biotechnol. 19:30–35. DOI 10.1016/j.copbio. 2007.11.006.
12. Stromberg, S., Bjorklund, M. G., Asplund, C., Skollermo, A., Persson, A., Wester, K., Kampf, C., Nilsson, P., Andersson, A. C., Uhlen, M., Kononen, J., Ponten, F., Asplund, A. (2007) A high-throughput strategy for protein profiling in cell microarrays using automated image analysis. Proteomics 7:2142–2150
13. Papay, J., Krenacs, T., Moldvay, J., Stelkovics, E., Furak, J., Molnar, B., Kopper, L. (2007) Immunophenotypic profiling of non-small cell lung cancer progression using the tissue microarray approach. Appl. Immunohistochem. Mol. Morphol. 15:19–30, 2007
14. Stelkovics, E., Korom, I., Marczinovits, I., Molnar, J., Rasky, K., Raso, E., Ficsor, E., Molnar, B., Kopper, L., Krenacs, T. (2008) Collagen XVII/BP180 protein expression in squamous cell carcinoma of the skin detected with novel monoclonal antibodies in archived tissues using tissue microarrays and digital microscopy. Appl. Immunohistochem. Mol. Morphol. 16:433–441
15. Anderson, W.F., Luo, S., Chatterjee, N., Rosenberg, P.S., Matsuno, R.K., Goodman, M.T., Hernandez, B.Y., Reichman, M., Dolled-Filhart, M.P., O'Regan, R.M., Garcia-Closas, M., Perou, C.M., Jatoi, I., Cartun, R,W., Sherman, M.E. (2008) Human epidermal growth factor receptor-2 and estrogen receptor expression, a demonstration project using the residual tissue repository of the Surveillance, Epidemiology, and End Results (SEER) program. Breast Cancer Res. Treat. 113:189–196
16. Turbin, D.A., Leung, S., Cheang, M.C., Kennecke, H.A., Montgomery, K.D., McKinney, S., Treaba, D.O., Boyd, N., Goldstein, L.C., Badve, S., Gown, A.M., van de Rijn, M., Nielsen, T.O., Gilks, C.B., Huntsman, D.G. (2007) Automated quantitative analysis of estrogen receptor expression in breast carcinoma does not differ from expert pathologist scoring: a tissue microarray study of 3,484 cases. Breast Cancer Res. Treat. 110:417–426
17. Rubin, M.A., Dunn, R., Strawderman, M., Pienta, K.J. (2002) Tissue microarray sampling strategy for prostate cancer biomarker analysis. Am. J. Surg. Pathol. 26:312–319.
18. Eisen, M.B., Spellman, P.T., Brown, P.O., Botstein, D, (1998) Cluster analysis and display of genome-wide expression patterns. Proc. Natl. Acad. Sci. USA 95:14863–14868
19. Liu, C.L., Prapong, W., Natkunam, Y., Alizadeh, A., Montgomery, K., Gilks, C.B., Rijn, M. (2002) Software tools for high-throughput analysis and archiving of immunohistochemistry staining data obtained with tissue microarrays. Am. J. Pathol. 161:1557–1565

Chapter 17

From Gene to Clinic: TMA-Based Clinical Validation of Molecular Markers in Prostate Cancer

Thorsten Schlomm, Felix KH Chun, and Andreas Erbersdobler

Abstract

Current high-throughput screening techniques using DNA arrays have identified hundreds of new candidate biomarkers for diagnosis and risk prediction of prostate cancer. Large-scale analysis of clinical prostate cancer specimens is a key prerequisite for the validation of these genes. We have constructed a tissue microarray from more than 2,500 prostate cancers with full histo-pathological and clinical long-term follow-up data and analyzed expression and gene copy number patterns of 16 different candidate markers for their ability to predict prostate cancer progression and patient prognosis. The best candidates were used to extend established clinical prediction tools (nomograms) that were based on nonmolecular data only, such as prostate-specific antigene (PSA), clinical stage, and histological grading (Gleason grade). Using this approach, we could identify ANXA3 as an independent marker, which was capable of increasing the accuracy of the clinical nomogram, thereby fulfilling the criteria of a novel prognostic prostate cancer marker. This approach of integrating large-scale clinical and molecular variables may provide a new paradigm for the use of molecular profiling to predict the clinical outcome in prostate cancer.

Key words: Prostate cancer, Tissue microarray, Nomogram, Translational research, Molecular marker

1. Introduction

Prostate cancer is the most common malignant tumor in men and the second leading cause of cancer death in western societies (1). Despite the high prevalence of the tumor, the clinical management of prostate cancer is limited by the low sensitivity and specificity of the existing diagnostic and prognostic tools.

Due to improved diagnostic screening techniques a high number of small organ-confined early prostate cancers and an increasing number of clinically insignificant tumors are diagnosed.

Ronald Simon (ed.), *Tissue Microarrays: Methods and Protocols*, Methods in Molecular Biology, vol. 664, DOI 10.1007/978-1-60761-806-5_17, © Springer Science+Business Media, LLC 2010

Patients with organ-confined cancers have a high likelihood of disease-free survival after standardized therapy (2–7). Therefore, it is important to predict which patient might benefit from an invasive therapy. In order to assess the most beneficial therapy, each patient has to be evaluated with regard to the expected therapy outcome. Currently, the most commonly applied prognostic markers for prostate cancer include clinical stage, histological tumor grade (Gleason score), and the serum marker PSA.

Many probability models (nomograms) have been developed on the basis of these clinico-pathological parameters to estimate the patient's risk to have a non-organ-confined tumor or lymph-node involvement (2–11). Patients with pathologically organ-confined prostate carcinomas have an approximately 90% chance of being cured by radical prostatectomy or radiation alone, whereas nearly all patients with lymph-node metastases will develop a biochemical progression within 10 years after surgery (2, 12–17). To date, the most accepted clinical prediction tools rely solely on the mentioned clinical, serological, and histological parameters.

In case of advanced metastasized cancers, androgen ablation (hormonal therapy) is the therapy of first choice, resulting in a transient reduction of symptoms until tumors become androgen-independent and then rapidly progress. Androgen-independent prostate cancer is largely resistant to chemotherapy. The success of new gene-specific drugs in other tumor entities raises the hope that prostate cancer patients might also benefit from similar targeted therapies in future.

Attempts toward improving prostate cancer patient management by molecular staging have not been successful so far. No single molecular parameter is routinely analyzed in prostate cancer tissue. This may be partly due to genuine properties of prostate cancer that may make this tumor a difficult target. Furthermore inherent logistical problems result in a shortage of prostate cancer tissue for research purposes.

Almost 10,000 scientific articles have been published on molecular features in prostate cancer between 1996 and 2008 according to Medline. Most of these studies suffered from small patient groups and do not link their data to clinical endpoints. A limited Medline search for the key words "prostate + cancer + immunohistochemistry + prognosis" and a review of these abstracts identified only 283 studies comparing specific molecular features with prognostic relevant clinical endpoints (Fig. 1) (18). Most of these studies included fewer than 100 patients, typically with heterogeneous treatment and diagnostic procedures of involved subjects. Typically, conflicting data were reported on the prognostic relevance of molecules investigated in more than one study (Table 1) and most biomarkers that were shown to be significantly associated or were of independent prognostic importance

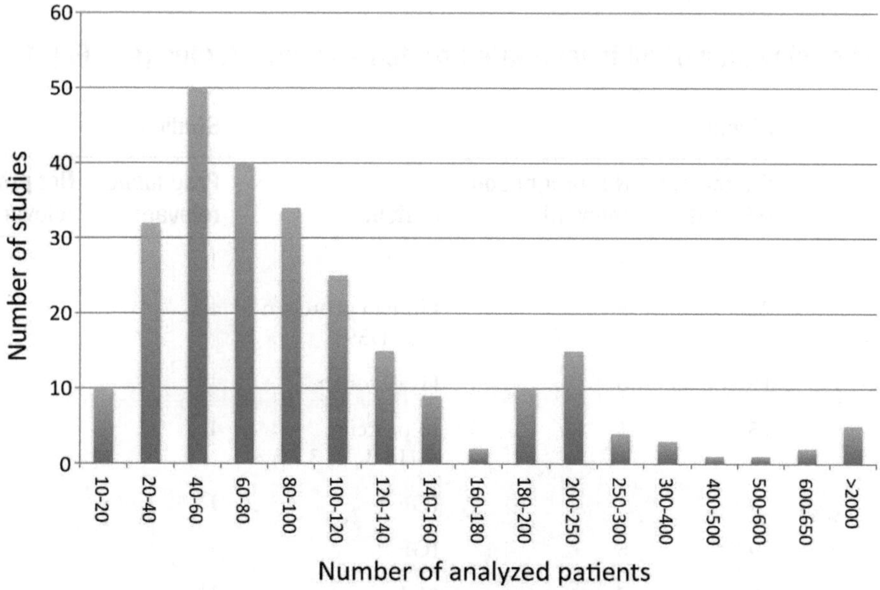

Fig. 1. Size of previous molecular studies in prostate cancer: Medline analysis ("prostate" + "cancer" + "immunohistochemistry" + "prognosis" + "year 1996–2008"). From over 4,000 listed articles, only 283 provide sufficient correlation to IHC results and PCA prognosis. Figure modified from Schlomm et al. (18).

in small pilot studies were subsequently not followed up in clinically significant large patient cohorts.

Considering that virtually no molecular marker has been successfully established for the daily care of prostate cancer yet, there is an urgent need for a powerful validation of all these new potential diagnostic and prognostic markers. The tissue microarray technology is optimally suited for the validation of molecular new candidate markers, provided that large sets of clinically well-defined tumor specimens with long-term clinical follow-up data are available. Optimally, promising molecular markers should be integrated into the existing prognosis prediction tools in order to increase the predictive power.

2. Clinical Prediction Tools in Prostate Cancer

There has been much progress in prostate cancer research in the last years, and clinicians have been provided with numerous tools like nomograms to assist with evidence-based medical decision-making (19–21). Nomograms represent risk stratification tools for patient counseling, which are capable of simultaneously considering multiple variables for individual risk assessment by combining either pre- or post-therapeutic information to predict a certain prostate cancer outcome. Accuracy represents one of the most important nomogram criteria. Current statistical methods offer the possibility of assessing a model's predictive accuracy. Predictive

Table 1
Prognostic relevance of all investigated proteins in 267 studies (see Fig. 1)

Protein	Studies		Protein	Studies	
	Prognostic relevant	Not prognostic relevant		Prognostic relevant	Not prognostic relevant
p53	29	5	hsp70	1	
Ki-67	22	5	Human protectin (CD59)	1	
Bcl-2	18	8	Hyaluronan	1	
AR	15	1	Id proteins (Id-1, 2, 3, 4)	1	
CD44	9		IGF-1	1	
Her-2	9	5	IGF-2	1	
p27	8	3	Mab	1	
CgA	7	5	MDM2	1	1
PSA	7	2	MIC-1	1	
alpha-catenin	5		MIF	1	
E-cadherin	5	1	MMPs	1	
NSE	5	2	MUC1	1	
Bax	4	3	Mucin1	1	
Beta-catenin	4		NEP	1	
p21	4	1	NPY	1	
Caveolin-1	3		Oncoprotein 18	1	
COX-2	3		Osteopontin	1	
NF kappa B	3		p120	1	
p16	3	1	P-Akt-1	1	
PCNA	3	1	Pepsinogen C	1	
serotonin (5HT)	3	1	PPP1CA	1	
survivin	3	2	S-100 protein	1	
FGF8	2		sFRP4	1	
hMSH2	2		Smad4	1	
IGFbp-3	2		Smad8	1	
MYC	2		Stat5	1	
Pin-1	2		Syndecan-1	1	
PLK1	2		TEF3	1	

(continued)

Table 1 (continued)

	Studies			Studies	
Protein	Prognostic relevant	Not prognostic relevant	Protein	Prognostic relevant	Not prognostic relevant
pRB	2		Tiam1	1	
PSMA	2		TSP-1	1	
TGF-beta1	2		VCP	1	
VEGF	2		vimentin	1	
actin	1		Wnt-1	1	
adrenomedullin	1		ZAG	1	
AKT/PKB	1		ADCP	1	
Akt-1, Akt-2, Akt-3	1		beta1C	1	
AMACR	1		bFGF	1	
AP-2	1		caspase-3		
AZGP1	1		CDK1	1	
BAG-1	1		CDK2	1	
BARK1	1		CDK6	1	
beta-MSP	1		CK 18	1	
BMP	1		Cyclin D3	1	
BMP2	1		delta-Catenin	1	
cathepsin D	1		DSPP	1	
CD24	1		GRK2	1	
CD31	1		hK2	1	
CD34	1		iNOS	1	
clusterin	1		NM23-H1	1	
cyclin D1	1	1	NSF	1	
desmin	1		p62	1	
EGFR	1	2	PPP2CB	1	
Endoglin	1		vWf		1
Ep-CAM	1	1	FHIT	1	
FAS	1	1	Her-3	1	
FGF17	1		hsp27	1	1

Table modified from Schlomm et al. (18)

accuracy should be ideally confirmed in an external cohort, which represents the "Gold-standard" method of validation. Alternatively, statistical methods such as bootstrapping may be used to internally validate the nomogram (19, 22–30). It is important to emphasize that predictive accuracy reflects two features at a time: discrimination and calibration. Conversely, the receiver operating characteristic (ROC) area under the curve (AUC) is only discriminatory. Usually, the discriminatory ability of predictive accuracy is derived from the ROC-AUC and expressed as a percentage. Discriminatory ability of predictive accuracy estimates ranges from 50 to 100%, where 50% is equivalent to a flip of a coin and 100% represents perfect prediction. The predictive accuracy of modern nomograms analyzing solely clinico-pathological variables is limited in a range between 70 and 80%. One of the major goals in clinical prostate cancer research is to increase the predictive accuracy of clinically useful prediction tools. Specifically, the emerging field of molecular biomarker research may substantially improve nomogram predictions. A novel marker should not only be judged according to its multivariable statistical significance, but should increase predictive accuracy of base predictors, in addition to confirming the independent, multivariable predictor status of this marker.

Such increases in predictive accuracy related to the use of nomograms may not only be of statistical significance but more importantly, clinically meaningful. For example, an increase in predictive accuracy of 12% translates into 120 men out of 1,000 patients who are provided with accurate predictions when the nomogram is used. This figure needs then to be extrapolated to the disease prevalence and subsequently to the number of diagnostic or therapeutic procedures. Thus, from a health, economic, medical, and personal standpoint, a relatively small increase in predictive accuracy translates into a clinically important number of patients who deserve to be provided with accurate predictions (19, 31).

Nomograms imperatively depend on their development cohorts. Most decision tools are based on cohorts of thousands of patients with accurate long time follow-up. For such retrospective patient cohorts, paraffin-embedded tissue represents the only source for molecular analyses. Therefore, Tissue Microarrays (TMAs) represent powerful tools to add molecular tumor biology information for potential improvement of established clinical nomograms.

3. Prostate Cancer TMA

To fit the needs for a TMA with sufficient statistical power to increase the predictive value of existing nomograms, we have constructed a tissue microarray from radical prostatectomy specimens from 3,261 patients, treated at the Department of Urology, University Medical Center Hamburg-Eppendorf between 1992 and 2003. Follow-up data were available for 2,385 patients, ranging from 1 to 144 months (mean, 34 months). None of the patients received adjuvant therapy. Additional (salvage) therapy was only initiated in case of a biochemical relapse. In all patients, PSA values were measured quarterly in the 1st year, followed by biannual measurements in the second and annual measurements after the 3rd year after surgery. Recurrence was defined as a postoperative PSA of 0.1 ng/ml and rising after an initial undetectable PSA. The first PSA value above or equal to 0.1 ng/ml was used to define the time of recurrence. Patients without evidence of tumor recurrence were censored at last follow-up.

All prostatectomy specimens were analyzed according to a standard procedure. All prostates were completely paraffin-embedded, including whole-mount sections (32). All Hematoxylin and Eosin (H&E)-stained histological sections from all prostatectomy specimens were reviewed for the purpose of this study and the index tumor, as defined by the largest tumor focus and/or the focus with the worst Gleason pattern, were marked on the slides. One 0.6 mm tissue core was punched out from the index tumor of each case, and transferred in a tissue microarray format (33). The 3,261 cores were distributed among seven tissue microarray blocks each containing 129–522 tumor samples. Each tissue microarray block also contained various control tissues including normal prostate tissue, other normal tissues and a set of tumor tissues. In addition, 37 lymph node metastases and 35 hormone refractory cancers were combined in a separate progression TMA. Hormone refractory prostate cancer was defined as: serum castration levels of testosterone, three consecutive rises of the prostatic-specific antigene (PSA) resulting in two 50% increases over the nadir, anti-androgen withdrawal for at least 4 weeks, PSA progression despite secondary hormonal manipulations or progression of osseous or soft tissue lesions (34).

The Hamburg prostate cancer collective is the cornerstone of numerous clinical nomograms (2, 3, 19, 35–50). Most tumors used to build these nomograms are included in our TMA. The validity of the clinical data attached to the arrayed tissue samples is demonstrated in an analysis comparing standard clinico-pathological parameters (Gleason grade, pT-stage, preoperative PSA serum level, presence of positive surgical margins) with biochemical tumor recurrence. All expected associations were found at a high level of statistical significance ($p < 0.0001$ each; Fig. 2).

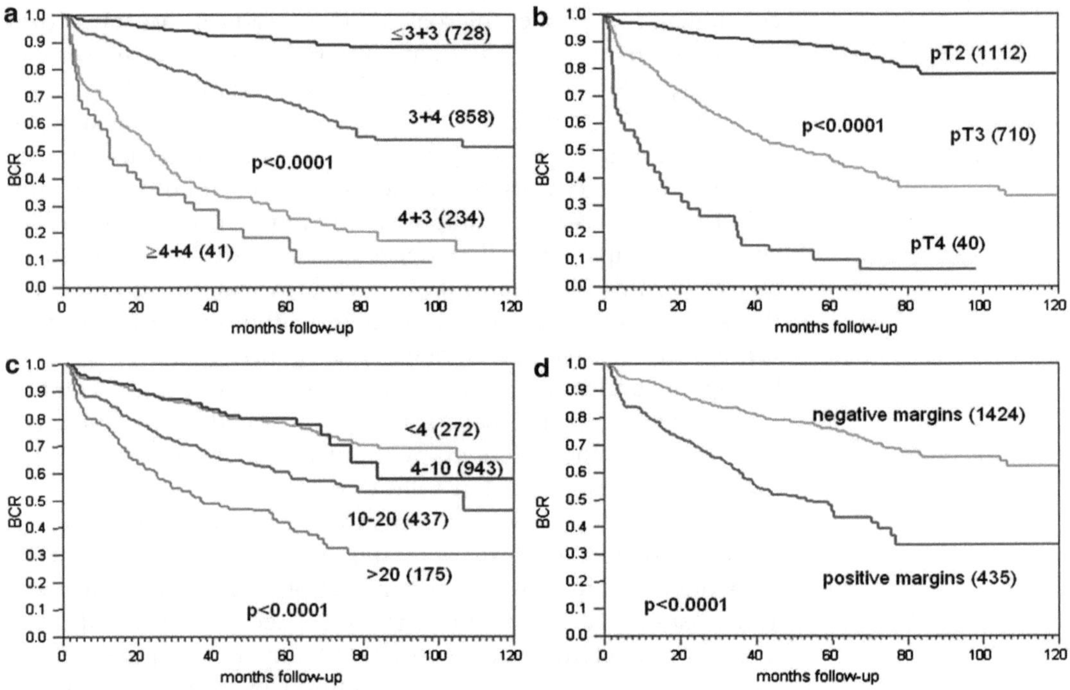

Fig. 2. Categories on the Hamburg prostate cancer TMA: Influence of clinico-pathological features on PSA recurrence. (**a**) Gleason grade, (**b**) pT category, (**c**) preoperative PSA, (**d**) Surgical margins. Figure modified from Schlomm et al. (55).

4. Transition of Molecular Data into Clinical Practice

Incorporation of emerging novel molecular markers into nomograms will potentially improve nomogram predictions. As outlined by Kattan et al., statistical significance is not synonymous with predictive ability. Instead, it is recommended that a novel marker should not only be judged according to its multivariable statistical significance, but should increase predictive accuracy of base predictors, in addition to confirming the independent, multivariable predictor status of this marker (19, 27, 31, 51, 52). In order to identify clinically useful biomarkers, we performed IHC and FISH analyzes of proteins and genes with well-known associations in prostate cancer progression. All analyzed marker showed a prognostic relevance in univariate or multivariate analyses (Table 2).

For example, we used our TMA to test the ability of Annexin A3 (ANXA3) to predict a PSA recurrence after radical prostatectomy. Complete clinical, pathological, and molecular data were available in 1,056 men. For analyses, pre-treatment PSA, digital rectal examination (DRE), and biopsy Gleason sum scores were used as base risk factors in univariable and multivariable Cox regression base models addressing the rate of PSA recurrence at 5 years after radical prostatectomy. TMA-derived information on

Table 2
Studies performed with the Hamburg Prostate Cancer TMA

Molecular target (protein/gene/ chromosome)	*n*	Positivity	Prognostic relevance (endpoint = BCR)	Reference
Ki 67 (labeling index)	2,612		Predictor (UV)	unpublished data
p53	2,514	2.5%	Independent predictor (MV)	(55)
PSA	2,612	98%	Predictor (UV)	unpublished data
PSMA	2,433	91.9%	Predictor (UV)	(56)
HER2 (IHC)	2,497	20%	Predictor (UV)	(57)
HER2 (FISH)	2,497	0.04% (amplification)		(57)
EGFR (IHC)	2,497	18%	Predictor (UV)	(58)
EGFR (FISH)	2,497	0.08% (amplification)		(58)
ANXA3	1,589	72.8%	Independent predictor (MV)	(59)
CD10	2,385	62.2%	Independent predictor (MV)	(60)
c-kit (CD117)	2,385	95.9%	Predictor (UV)	(61)
NF-κB (p65)	1,161	94%	Independent predictor (MV)	(62)
Chromogranin A	2,398	17.2%	Predictor (UV)	(63)
Synaptophysin	2,398	11.2%	Predictor (UV)	(63)
Chromosome 8p (FISH)	1,292	31.86% (loss)	Predictor (UV)	(64)
Chromosome 8q (FISH)	1,292	8.19% (gain)	Predictor (UV)	(64)

BCR, biochemical recurrence; *UV*, univariate analysis; *MV*, multivariate analysis

ANXA3 coded as negative, weak, moderate, and strong complemented the multivariable models (53). Cox regression coefficients were then used to construct prognostic nomograms to predict the PSA recurrence free rate at 5 years after radical prostatectomy. The predictive accuracy of the nomograms was quantified using Harrell's concordance index. Accuracy of predictions was quantified with and without ANXA3. This method was selected with the intent of quantifying the increment in predictive accuracy, associated with the addition of ANXA3 to established predictors. Two hundred bootstrap re-samples were used to reduce over fit bias. All statistical tests were performed using S-PLUS Professional, version 1 (MathSoft

Inc., Seattle, Washington). Moreover, all tests were two-sided with a significance level at 0.05. In univariable and multivariable Cox regression models, all risk factors including ANXA3 were statistically, significantly, and independently associated with a PSA recurrence (all $p \leq 0.041$). Five years after radical prostatectomy, the predictive accuracy of multivariable Cox regression models addressing PSA-free recurrence increased from 0.71 to 0.73 when ANXA3 was added to the established base risk factors (Fig. 3).

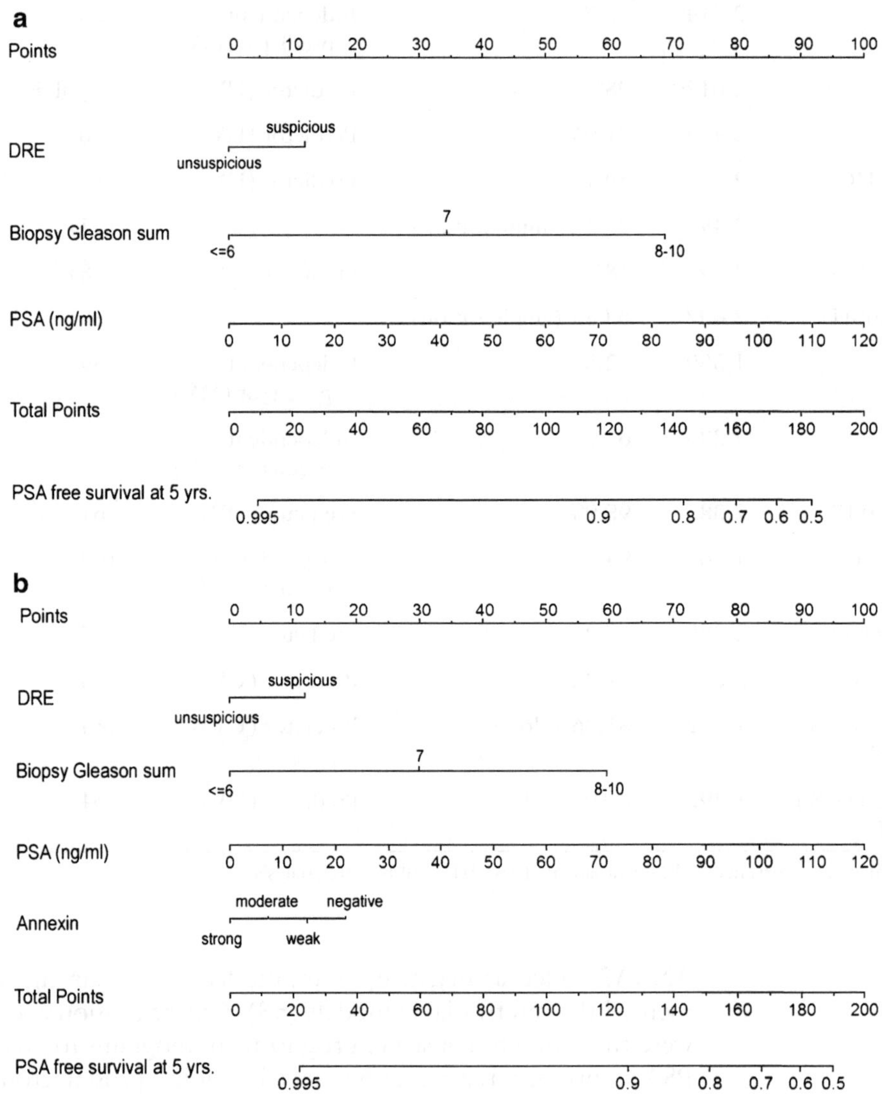

Fig. 3. Nomogram predicting biochemical recurrence (PSA relapse) after radical prostatectomy: (**a**) Nomogram with clinico-pathological data. (**b**) Same Nomogram combined with Annexin A3.Locate patient PSA on PSA (ng/ml) axis. Draw line straight up to point axis to determine how many points patient receives for probability of BCR after radical prostatectomy. Repeat this process for all other predictors, each time drawing straight upward line to point axis. Sum points for each predictor and locate this sum on total point axis. Draw line straight down to find patient probability of a BCR after radical prostaectomy. Inclusion of Annexin A3 improved the predictive accuracy of the nomogram.

5. Prostate Cancer TMA as a Model of Prostate Biopsies

The higher likelihood of positive immunostainings (false and true) in larger sized tissue samples makes it difficult to use prognostic associations of molecular features identified in tissue microarray studies for clinical routine (54). Extensive validation experiments and, potentially, also adjustments of protocols and criteria for "positivity" may be required before routine use affects clinical decision making. It is important to note that in the case of prostate cancer, there may be an exception from this rule. In this cancer, initial diagnosis is exclusively made on very small tissue samples (biopsies), for which tissue microarrays may represent an ideal model for routine molecular analysis simulation. The amount of tumor available for molecular analysis on needle core biopsies is approximately comparable to the situation on tissue microarrays. Therefore, it can be speculated that prognostic biomarkers identified on tissue microarrays may be transferable to needle core biopsies and thus be utilized for an improved preoperative risk assessment with potential importance on the therapeutic decision-making.

References

1. Ferlay J, Autier P, Boniol M, Heanue M, Colombet M, Boyle P. Estimates of the cancer incidence and mortality in Europe in 2006. Ann Oncol 2007;18:581–92.
2. Graefen M, Augustin H, Karakiewicz PI, et al. Can predictive models for prostate cancer patients derived in the United States of America be utilized in European patients? A validation study of the Partin tables. Eur Urol 2003;43:6–10; discussion 1.
3. Graefen M, Noldus J, Pichlmeier U, et al. Early prostate-specific antigen relapse after radical retropubic prostatectomy: prediction on the basis of preoperative and postoperative tumor characteristics. Eur Urol 1999;36:21–30.
4. Kattan MW, Eastham JA, Stapleton AM, Wheeler TM, Scardino PT. A preoperative nomogram for disease recurrence following radical prostatectomy for prostate cancer. J Natl Cancer Inst 1998;90:766–71.
5. Kattan MW, Potters L, Blasko JC, et al. Pretreatment nomogram for predicting freedom from recurrence after permanent prostate brachytherapy in prostate cancer. Urology 2001;58:393–9.
6. Kattan MW, Zelefsky MJ, Kupelian PA, Scardino PT, Fuks Z, Leibel SA. Pretreatment nomogram for predicting the outcome of three-dimensional conformal radiotherapy in prostate cancer. J Clin Oncol 2000;18:3352–9.
7. Partin AW, Kattan MW, Subong EN, et al. Combination of prostate-specific antigen, clinical stage, and Gleason score to predict pathological stage of localized prostate cancer. A multi-institutional update. Jama 1997;277:1445–51.
8. Bauer JJ, Connelly RR, Seterhenn IA, et al. Biostatistical modeling using traditional preoperative and pathological prognostic variables in the selection of men at high risk for disease recurrence after radical prostatectomy for prostate cancer. J Urol 1998;159:929–33.
9. D'Amico AV, Whittington R, Malkowicz SB, et al. Pretreatment nomogram for prostate-specific antigen recurrence after radical prostatectomy or external-beam radiation therapy for clinically localized prostate cancer. J Clin Oncol 1999;17:168–72.
10. Djavan B, Remzi M, Zlotta A, Seitz C, Snow P, Marberger M. Novel artificial neural network for early detection of prostate cancer. J Clin Oncol 2002;20:921–9.
11. Stamey TA, McNeal JE, Yemoto CM, Sigal BM, Johnstone IM. Biological determinants of cancer progression in men with prostate cancer. Jama 1999;281:1395–400.
12. Catalona WJ, Smith DS. 5-year tumor recurrence rates after anatomical radical retropubic

prostatectomy for prostate cancer. J Urol 1994;152:1837–42.

13. Epstein JI, Partin AW, Potter SR, Walsh PC. Adenocarcinoma of the prostate invading the seminal vesicle: prognostic stratification based on pathologic parameters. Urology 2000; 56:283–8.
14. Epstein JI, Pizov G, Walsh PC. Correlation of pathologic findings with progression after radical retropubic prostatectomy. Cancer 1993a;71:3582–93.
15. Gibbons RP, Correa RJ, Jr., Brannen GE, Weissman RM. Total prostatectomy for clinically localized prostatic cancer: long-term results. J Urol 1989;141:564–6.
16. Pound CR, Partin AW, Epstein JI, Walsh PC. Prostate-specific antigen after anatomic radical retropubic prostatectomy. Patterns of recurrence and cancer control. Urol Clin North Am 1997;24:395–406.
17. Zincke H, Oesterling JE, Blute ML, Bergstralh EJ, Myers RP, Barrett DM. Long-term (15 years) results after radical prostatectomy for clinically localized (stage T2c or lower) prostate cancer. J Urol 1994;152:1850–7.
18. Schlomm T, Erbersdobler A, Mirlacher M, Sauter G. Molecular staging of prostate cancer in the year 2007. World J Urol 2007;25:19–30.
19. Chun FK, Karakiewicz PI, Briganti A, et al. Prostate cancer nomograms: an update. Eur Urol 2006;50:914–26; discussion 26.
20. Karakiewicz PI, Briganti A, Chun FK, Valiquette L. Outcomes Research: A Methodologic Review. Eur Urol 2006.
21. Ross PL, Scardino PT, Kattan MW. A catalog of prostate cancer nomograms. J Urol 2001; 165:1562–8.
22. Bradley E, Tibshirani RJ. Monographs on statistics and applied probability: An introduction to the bootstrap. Champman and Hall/ CRC 1993:275.
23. Eastham JA, Kattan MW, Scardino PT. Nomograms as predictive models. Semin Urol Oncol 2002;20:108–15.
24. Karakiewicz PI, Chun FK, Briganti A, et al. Prostate cancer nomograms are superior to neural networks. Can J Urol 2006;13 Suppl 2:18–25.
25. Kattan M. Statistical prediction models, artificial neural networks, and the sophism "I am a patient, not a statistic". J Clin Oncol 2002;20: 885–7.
26. Kattan MW. Nomograms. Introduction. Semin Urol Oncol 2002;20:79–81.
27. Kattan MW. Judging new markers by their ability to improve predictive accuracy. J Natl Cancer Inst 2003;95:634–5.
28. Kattan MW. Comparison of Cox regression with other methods for determining prediction models and nomograms. J Urol 2003;170:S6–9; discussion S10.
29. Steyerberg EW, Harrell FE, Jr., Borsboom GJ, Eijkemans MJ, Vergouwe Y, Habbema JD. Internal validation of predictive models: efficiency of some procedures for logistic regression analysis. J Clin Epidemiol 2001;54:774–81.
30. Steyerberg EW, Harrell FE, Jr., Goodman PH. Neural networks, logistic regression, and calibration. Med Decis Making 1998;18:349–50.
31. Specht MC, Kattan MW, Gonen M, Fey J, Van Zee KJ. Predicting nonsentinel node status after positive sentinel lymph biopsy for breast cancer: clinicians versus nomogram. Ann Surg Oncol 2005;12:654–9.
32. Erbersdobler A, Fritz H, Schnoger S, et al. Tumour grade, proliferation, apoptosis, microvessel density, p53, and bcl-2 in prostate cancers: differences between tumours located in the transition zone and in the peripheral zone. Eur Urol 2002;41:40–6.
33. Bubendorf L, Kononen J, Koivisto P, et al. Survey of gene amplifications during prostate cancer progression by high-throughout fluorescence in situ hybridization on tissue microarrays. Cancer Res 1999;59:803–6.
34. Heidenreich A, Aus G, Bolla M, et al. EAU Guidelines on prostate cancer. Eur Urol 2008;53:68–80.
35. Graefen M, Karakiewicz PI, Cagiannos I, et al. Validation study of the accuracy of a postoperative nomogram for recurrence after radical prostatectomy for localized prostate cancer. J Clin Oncol 2002;20:951–6.
36. Graefen M, Karakiewicz PI, Cagiannos I, et al. International validation of a preoperative nomogram for prostate cancer recurrence after radical prostatectomy. J Clin Oncol 2002;20:3206–12.
37. Suardi N, Porter CR, Reuther AM, et al. A nomogram predicting long-term biochemical recurrence after radical prostatectomy. Cancer 2008;112:1254–63.
38. Gallina A, Chun FK, Briganti A, et al. Development and split-sample validation of a nomogram predicting the probability of seminal vesicle invasion at radical prostatectomy. Eur Urol 2007;52:98–105.
39. Chun FK, Briganti A, Jeldres C, et al. Tumour volume and high grade tumour volume are the best predictors of pathologic stage and biochemical recurrence after radical prostatectomy. Eur J Cancer 2007;43:536–43.
40. Steuber T, Chun FK, Erbersdobler A, et al. Development and internal validation of preoperative transition zone prostate cancer nomogram. Urology 2006;68: 1295–300.

41. Steuber T, Graefen M, Haese A, et al. Validation of a nomogram for prediction of side specific extracapsular extension at radical prostatectomy. J Urol 2006;175:939–44; discussion 44.
42. Chun FK, Steuber T, Erbersdobler A, et al. Development and internal validation of a nomogram predicting the probability of prostate cancer Gleason sum upgrading between biopsy and radical prostatectomy pathology. Eur Urol 2006;49:820–6.
43. Augustin H, Eggert T, Wenske S, et al. Comparison of accuracy between the Partin tables of 1997 and 2001 to predict final pathological stage in clinically localized prostate cancer. J Urol 2004;171:177–81.
44. Cagiannos I, Karakiewicz P, Eastham JA, et al. A preoperative nomogram identifying decreased risk of positive pelvic lymph nodes in patients with prostate cancer. J Urol 2003;170:1798–803.
45. Kattan MW, Eastham JA, Wheeler TM, et al. Counseling men with prostate cancer: a nomogram for predicting the presence of small, moderately differentiated, confined tumors. J Urol 2003;170:1792–7.
46. Augustin H, Hammerer PG, Graefen M, et al. Insignificant prostate cancer in radical prostatectomy specimen: time trends and preoperative prediction. Eur Urol 2003;43:455–60.
47. Graefen M, Karakiewicz PI, Cagiannos I, et al. A validation of two preoperative nomograms predicting recurrence following radical prostatectomy in a cohort of European men. Urol Oncol 2002;7:141–6.
48. Palisaar RJ, Graefen M, Karakiewicz PI, et al. Assessment of clinical and pathologic characteristics predisposing to disease recurrence following radical prostatectomy in men with pathologically organ-confined prostate cancer. Eur Urol 2002;41:155–61.
49. Conrad S, Graefen M, Pichlmeier U, et al. Prospective validation of an algorithm with systematic sextant biopsy to predict pelvic lymph node metastasis in patients with clinically localized prostatic carcinoma. J Urol 2002;167:521–5.
50. Graefen M, Haese A, Pichlmeier U, et al. A validated strategy for side specific prediction of organ confined prostate cancer: a tool to select for nerve sparing radical prostatectomy. J Urol 2001;165:857–63.
51. Kattan MW. Evaluating a new marker's predictive contribution. Clin Cancer Res 2004; 10:822–4.
52. Katz EM, Kattan MW. How to judge a tumor marker. Nat Clin Pract Oncol 2005;2:482–3.
53. Kollermann J, Schlomm T, Bang H, et al. Expression and Prognostic Relevance of Annexin A3 in Prostate Cancer. Eur Urol 2008.
54. Torhorst J, Bucher C, Kononen J, et al. Tissue microarrays for rapid linking of molecular changes to clinical endpoints. Am J Pathol 2001;159:2249–56.
55. Schlomm T, Iwers L, Kirstein P, et al. Clinical significance of p53 alterations in surgically treated prostate cancers. Mod Pathol 2008;21:1371–1378.
56. Schlomm T, Petersen A, Kramme I, et al. Alterations of the physiological PSMA expression patterns are associated with poor prognosis in prostate cancer. (submitted).
57. Minner S, Jessen B, Stiedenroth L, et al. Low level HER2 overexpression is associated with rapid tumor cell proliferation and poor prognosis in prostate cancer. Clin Cancer Res 2010;16:1553–1560.
58. Schlomm T, Kirstein P, Iwers L, et al. Clinical significance of epidermal growth factor receptor protein overexpression and gene copy number gains in prostate cancer. Clin Cancer Res 2007;13:6579–84.
59. Kollermann J, Schlomm T, Bang H, et al. Expression and prognostic relevance of annexin a3 in prostate cancer. Eur Urol 2008; 54:1314–23.
60. Fleischmann A, Schlomm T, Huland H, et al. Distinct subcellular expression patterns of neutral endopeptidase (CD10) in prostate cancer predict diverging clinical courses in surgically treated patients. Clin Cancer Res 2008;14:7838–42.
61. Fleischmann A, Schlomm T, Köllermnn J, et al. Immunological microenvironment in prostate cancer: high mast cell densities are associated with favorable tumor characteristics and good prognosis. Prostate 2009;69:976–981.
62. Lessard L, Bégin LR, Schlomm T, et al. Validation of NF-kappaB p65 as a prostate cancer prognostic marker on a large cohort of European patients. J Clin Oncol 2008;26: abstr.11055.
63. Currlin E, Eichelberg C, Graefen M, et al. Neuroendokrine Differenzierung des Prostatakarzinoms korreliert mit erhöhter Tumorprogression. Jahrestagung der Vereinigung Norddeutscher Urologen (VNU) 2007.
64. El Gammal AT, Brüchmann M, Zustin J, et al. Chromosome 8p deletions and 8q gains are associated with tumor progression and poor prognosis in prostate cancer. Clin Cancer Res 2010;16:56–54.

Index

Ronald Simon (ed.), *Tissue Microarrays: Methods and Protocols*, Methods in Molecular Biology, vol. 664,
DOI 10.1007/978-1-60761-806-5, © Springer Science+Business Media, LLC 2010

MIX
Papier aus verantwortungsvollen Quellen
Paper from responsible sources
FSC® C105338
FSC
www.fsc.org

If you have any concerns about our products,
you can contact us on
ProductSafety@springernature.com

In case Publisher is established outside the EU,
the EU authorized representative is:
Springer Nature Customer Service Center GmbH
Europaplatz 3, 69115 Heidelberg, Germany

Printed by Libri Plureos GmbH
in Hamburg, Germany